# Global Perspectives on Health Geography

**Series Editor**
Valorie Crooks, Department of Geography, Simon Fraser University,
Burnaby, BC, Canada

Global Perspectives on Health Geography showcases cutting-edge health geography research that addresses pressing, contemporary aspects of the health-place interface. The bi-directional influence between health and place has been acknowledged for centuries, and understanding traditional and contemporary aspects of this connection is at the core of the discipline of health geography. Health geographers, for example, have: shown the complex ways in which places influence and directly impact our health; documented how and why we seek specific spaces to improve our wellbeing; and revealed how policies and practices across multiple scales affect health care delivery and receipt.

The series publishes a comprehensive portfolio of monographs and edited volumes that document the latest research in this important discipline. Proposals are accepted across a broad and ever-developing swath of topics as diverse as the discipline of health geography itself, including transnational health mobilities, experiential accounts of health and wellbeing, global-local health policies and practices, mHealth, environmental health (in)equity, theoretical approaches, and emerging spatial technologies as they relate to health and health services. Volumes in this series draw forth new methods, ways of thinking, and approaches to examining spatial and place-based aspects of health and health care across scales. They also weave together connections between health geography and other health and social science disciplines, and in doing so highlight the importance of spatial thinking.

Dr. Valorie Crooks (Simon Fraser University, crooks@sfu.ca) is the Series Editor of Global Perspectives on Health Geography. An author/editor questionnaire and book proposal form can be obtained from Publishing Editor Zachary Romano (zachary.romano@springer.com).

More information about this series at https://link.springer.com/bookseries/15801

Caroline Barakat • Rania Dghaim
Fatme Al Anouti
Editors

# Adolescent Health in the Middle East and North Africa

## An Epidemiological Perspective

Springer

*Editors*
Caroline Barakat
University of Ontario Institute of Technology
Oshawa, ON, Canada

Rania Dghaim
Zayed University
Dubai, United Arab Emirates

Fatme Al Anouti
Zayed University
Abu Dhabi, United Arab Emirates

ISSN 2522-8005 ISSN 2522-8013 (electronic)
Global Perspectives on Health Geography
ISBN 978-3-030-92106-4 ISBN 978-3-030-92107-1 (eBook)
https://doi.org/10.1007/978-3-030-92107-1



This Springer imprint is published by the registered company Springer Nature Switzerland AG
The registered company address is: Gewerbestrasse 11, 6330 Cham, Switzerland

# Acknowledgements

We would like to thank all researchers in the MENA region who continue to demonstrate deep enthusiasm and commitment towards adolescent health and well-being. We dedicate this book to our children and all adolescents, especially those from the Middle East and North Africa. We hope that our research helps shed light on your needs and your journey.

# Contents

# Contributors

**Fatme Al Anouti** Department of Natural Sciences and Public Health, College of Natural and Health Sciences, Zayed University, Abu Dhabi, United Arab Emirates

**Ayesha Salem Al Dhaheri** Department of Nutrition and Health, College of Medicine and Health Sciences, United Arab Emirates University, Al Ain, United Arab Emirates

**Moubadda Assi** School of Social Psychology, Nottingham Trent University, London, UK

National AIDS Program Lebanon, Beirut, Lebanon

**Caroline Barakat** Faculty of Health Sciences, Ontario Tech University, Oshawa, ON, Canada

**Rania Dghaim** Department of Natural Sciences and Public Health, College of Natural and Health Sciences, Zayed University, Dubai, United Arab Emirates

**Nour El Asswad** Department of Biology, University of Toronto, Toronto, ON, Canada

**Reham Khaled Abou El Fadl** Faculty of Dentistry, Ain Shams University, Cairo, Egypt

MPH Imperial College London, London, UK

**Mirna Fawaz** Nursing Department, Faculty of Health Sciences, Beirut Arab University, Beirut, Lebanon

**Leila Cheikh Ismail** Department of Clinical Nutrition and Dietetics, College of Health Sciences, University of Sharjah, Sharjah, United Arab Emirates

**Moatassem Kenaan** Faculty of Health Sciences, Ontario Tech University, Oshawa, ON, Canada

**Kinda Khalaf** Department of Biomedical Engineering, Health Engineering Innovation Center, Khalifa University of Science and Technology, Abu Dhabi, United Arab Emirates

**Ismaël Maatouk** Dermatology Department, Clemenceau Medical Center affiliated with Johns Hopkins, Beirut, Lebanon

School of Social Psychology, Nottingham Trent University, London, UK

**Samah Mohammed** Faculty of Health Sciences, Ontario Tech University, Oshawa, ON, Canada

**Dhanya Menoth Mohan** Department of Biomedical Engineering, Health Engineering Innovation Center, Khalifa University of Science and Technology, Abu Dhabi, United Arab Emirates

**Haneen Raafat Fathi Mousa** Faculty of Dentistry, Ain Shams University, Cairo, Egypt

**Shatha Muhsineh** Dubai Health Authority, Dubai, United Arab Emirates

**Fatma Nasser** College of Natural and Health Sciences, Zayed University, Dubai, United Arab Emirates

**Doaa Oraby** Independent Reproductive Health Consultant, Cairo, Egypt

**Susan Yousufzai** Faculty of Health Sciences, Ontario Tech University, Oshawa, ON, Canada

# Part I
# Introduction

# Chapter 1
# Adolescent Health in the Middle East and North Africa: Determinants and Distribution

**Caroline Barakat, Moatassem Kenaan, Fatme Al Anouti, and Rania Dghaim**

Adolescence is an important phase of growth characterized as the transition period from childhood into adulthood. The United Nations defines an adolescent as an individual between the ages of 10 and 19 years. Adolescents make up 16% of the world's population (1.2 billion) (Bhutta et al. 2019). Adolescents are vulnerable to a myriad of factors that influence their growth and development (Patel et al. 2008). Changes in hormones, emotions, and behaviors characterize the accelerated development that adolescents undergo and influence their health (Currie et al. 2008). In addition, rapid development and changes that occur influence how an adolescent perceives the world around him or her. Moreover, individual physical and social development during adolescence influences one's future health outcomes (Bhutta et al. 2019).

Abstract thinking, development of sexual identity, and the increased ability to reason are psychological and psychosocial changes that occur during puberty (Currie et al. 2008). An increase in cognitive and intellectual capacity occurs during

Caroline Barakat and Moatassem Kenaan contributed equally with all other contributors.

C. Barakat (✉) · M. Kenaan
Faculty of Health Sciences, Ontario Tech University, Oshawa, ON, Canada
e-mail: Caroline.Barakat@ontariotechu.ca; moatassem.kenaan@ontariotechu.ca

F. Al Anouti
Department of Natural Sciences and Public Health, Zayed University, Abu Dhabi, United Arab Emirates

R. Dghaim
Department of Natural Sciences and Public Health, Zayed University, Dubai, United Arab Emirates



C. Barakat et al. (eds.), *Adolescent Health in the Middle East and North Africa*, Global Perspectives on Health Geography,
https://doi.org/10.1007/978-3-030-92107-1_1

the neurodevelopmental stages of adolescence (Bhutta et al. 2019). External pressures, such as social media, schools, and peer groups, can profoundly influence adolescents' health. Other factors include social stigma, socioeconomic status, poverty, violence, education, and social support (Currie et al. 2008). Health is a complex topic influenced by an inner web of different factors. Socioeconomic status is a major determinant of adolescent health as it influences their education, diet, and environment. Healthy dietary behaviors during adolescence play a vital role in chronic and acute disease development during later stages in life (Currie et al. 2008). The prefrontal cortex, which develops during adolescence, is an area in the brain that is responsible for planning for the future and decision-making. Proper education can stimulate the full development of the adolescent brain, allowing them to reach their maximum potential (Patel et al. 2008). Considering the various biological and physiological changes during this period, it is essential to assess and monitor adolescent health and the related determinants of adolescent health.

While the transition period into adulthood is universal, given that interactions within one's environment shape health and quality of life, experiences among adolescents from various parts of the world are in no way uniform. *Adolescent Health in the Middle East and North Africa: An Epidemiological Perspective* is the first edition of an editorial book that examines various aspects of adolescent health in the Middle East and North Africa (MENA). The book consists of two sections: adolescent determinants of health and adolescent health status. It examines adolescent health outcomes related to oral health, diabetes, nutrition, communicable diseases, reproductive health, respiratory health, and physical activity. Although the adolescent population is rapidly increasing in the MENA region, this subpopulation is commonly neglected and is often marginalized (AlBuhairan 2015). Thus, *Adolescent Health in the Middle East and North Africa: An Epidemiological Perspective* will also touch upon policy changes and programs that if adopted could have the capacity to improve the lives of MENA adolescents.

MENA countries include Palestine, Jordan, Lebanon, Morocco, Libya, Algeria, Egypt, Iran, Iraq, Bahrain, Kuwait, the United Arab Emirates (UAE), the Kingdom of Saudi Arabia (KSA), Qatar, Syria, Yemen, and Tunisia (Fig. 1.1). The MENA region is unique as many nations suffer from war conflicts, which elicit high rates of refugees and families living under the poverty line (AlBuhairan 2015). A large proportion of the population are exposed to various physical and environmental risk factors that impact their health, such as poor housing, nutrition, and lack of opportunities for physical activity. Some nations have massive disparities between men and women, leading to social inequality between genders through allocating more resources to adolescent boys while neglecting girls (AlBuhairan 2015).

The first five chapters of this book focus on select determinants of adolescent health. The first chapter will discuss the nutritional habits of adolescents, the different factors that influence their eating behaviors, and the various eating disorders that adolescents can develop. Information related to the social determinants that affect adolescent oral health and tobacco use in relation to MENA adolescents follow. The third chapter focuses on physical activity and sedentary behaviors among MENA adolescents. The next chapter discusses obesity predictive modelling based on a list

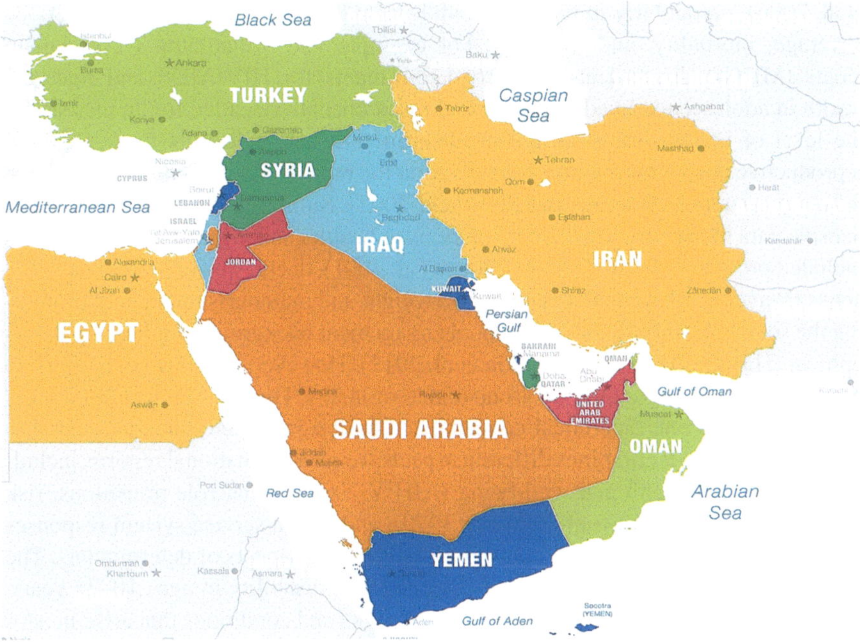


**Fig. 1.1** Map of MENA. (Source: https://www.worldatlas.com)

of risk factors and highlights multifactorial aspects of adolescent and childhood obesity in the MENA region.

The second section of this book focuses on adolescent health outcomes. Cheikh Ismail and Al Dhaheri provide an overview of the current literature on the standardized measures of adolescent growth, with a focus on adolescent growth assessment in MENA. Abou El Fadl and Fathi Mousa focus on the prevalence of common oral problems, such as fluorosis and dental caries among adolescents in the MENA region. Dghaim et al. discuss respiratory and allergic disorders among adolescents in the UAE, followed by Yousufzai and Barakat's chapter on diabetes mellitus in the MENA region. The final chapter provides an epidemiological overview of HIV and adolescent sexual health in MENA.

## 1.1 Adolescent Health Determinants

Health determinants influence adolescent health and health risks. Health determinants are usually established within the framework of initiating a particular policy, such as disease-associated initiatives. This approach has resulted in a large increase in the number of determinants, which has challenged worldwide health information systems (Murray 2007). Many determinants remain inadequately defined, which raises the need to describe a number of fundamental health determinants (Bchir et al. 2006), particularly in relation to adolescent health risk factors, health service coverage, morbidity, and mortality. For example, the Millennium Development Goals (MDGs) directed attention to determinants for HIV/AIDS and maternal health in adolescents. In addition, several researchers have called for an increase in the level of knowledge pertaining to determinants and indicators for sexual and reproductive health, taking into consideration the rapid shifts in adolescent health in different countries (Boerma and Stansfield 2007). Similarly, the need for comprehensive data to encourage efficient solutions to health challenges of adolescents has included preventing traffic injury (Shope et al. 2001), decreasing adolescent alcohol abuse (Wagenaar 2003), addressing underweight and malnourishment, and promoting the social, neighborhood, and school engagement background for healthy development (Hawkins et al. 2009; Patton et al. 2011). However, to this time, there is no present worldwide agreed-upon group of determinants. The *2012 Lancet Adolescent Health Series* is one theoretical concept used to designate determinants related to adolescent health. It combines different aspects from earlier national reports, including measures of health and well-being (AIHW 2007), social role transitions, risk and protective factors (Bronfenbrenner 1979), and health service system responses (Rosen and Levine 2010). The concept highlights five aspects of determinants. The first aspect focuses on key causes of mortality and disability in ages 10–24 years. The second aspect reflects health-related behaviors and conditions that arise in adolescence and increase risks for diseases (both present and future). The third aspect emphasizes risk and protective factors related to social circumstances that influence health risks. The fourth aspect recognizes indicators of social role shifts related to different forms of health risk. Finally, the fifth aspect focuses on health service policy interventions that have the capacity to affect present or future health status of adolescents. Four types of databases form the basis of this concept, including the global mortality database, international household surveys, international school-based surveys, and other United Nations (UN) sources (Murray and Lopez 2013).

Recognizing that an array of environmental, social, economic, and personal factors determine health, the intricate relationships among all factors that determine adolescents' health naturally extend beyond the boundaries of traditional health care. Thus, determinants of adolescent health include several structural factors, such as access to education, income inequality, and various physical and social environmental factors. The following is an overview of select important determinants of health during adolescence.

### 1.1.1 Income Inequality and Socioeconomic Status (SES)

An individual's physical and psychological health is correlated to his or her socioeconomic status (SES). In general, people with higher SES have relatively better health outcomes and lower mortality and morbidity rates than those with lower SES (Cameron and Williams 2009). Research documents positive associations between familial SES and satisfaction with health, resilience, comfort, and avoidance of risks among children and adolescents (Starfield et al. 2002). Adolescents in low-income families experience problems in reasoning and linguistic abilities (Najman et al. 2004). On the other hand, adolescents who have access to cultural, social, and educational services and support through their families and experience for cultures (e.g., family holidays which allow them to experience cultural diversity) have a relatively higher health-related quality of life (Von Rueden et al. 2006). SES can also affect access to and quality of education, as families with relatively higher SES are able to enroll their children in better schools, while families with low SES may be restricted to enrolling their children in public schools.

### *1.1.2 Obesity and Nutritional Habits*

Obesity affects two billion individuals and approximately 340 million children and adolescents worldwide (The World Health Organization [WHO] 2020). A high BMI score is associated with an increased risk of type 2 diabetes, cardiovascular diseases, sleep apnea, and hypertension (Christian and Smith 2018). In the last 30 years, obesity rates have doubled among adolescents; several factors contribute to this increase, such as high caloric intake and a sedentary lifestyle. Countries in the MENA region are at great risk of increasing adolescent obesity due to recent westernization and socioeconomic advances. In the UAE, obesity places a $6 billion economic burden associated with disease costs. Emirati children are 1.8 times more obese than American children are, and the disparity between the two populations is on the rise. The WHO (2020) has expressed great concern over the increasing rates of obesity in MENA adolescents. Therefore, prevention programs and intervention strategies should be adopted in the MENA region to tackle the obesity pandemic.

Several factors like poverty and lower information on healthy food choices hinder adolescents from accessing healthy food. This is particularly important for adolescents during their growth phases, as they require adequate body weight to become healthy adults (Pal et al. 2017). According to Christian and Smith (2018), being underweight or overweight during adolescence can lead to several health complications. Underweight adolescents likely have poor nutritional intake and micronutrient deficiencies, which in turn raises the risk for negative educational, behavioral, and cognitive outcomes (Galler et al. 2017). Therefore, the global burden of micronutrient deficiencies in adolescents is high in countries with relatively lower Social Development Index (SDI) (Akseer et al. 2017). The most common nutritional deficiencies among adolescents are iron and vitamin A (Christian and Smith 2018). Iron deficiency is highly prevalent among adolescent girls living in low-middle SDI countries, while vitamin A deficiency is highly prevalent among adolescent boys living in low SDI countries (Christian and Smith 2018).

### 1.1.3 Access to Education

One of the Millennium Development Goals requires all children to complete their primary school worldwide in order to secure better future and health (Viner et al. 2012). Not only does education contribute to improving health, education also empowers adolescents by allowing them to have more responsibility and control of their own lives (Gakidou et al. 2010; Kravdal 2002). Access to adolescent education differs in countries according to income. In this regard, an estimate of more than 33% of adolescents do not have access to schools in low-income countries as compared to 4% in high-income countries (Bruneforth and Wallet 2010). Furthermore, education can protect adolescents against a full range of health risk behaviors, particularly for those who reside in middle-income and low-income countries (Anteghini et al. 2001; Blum et al. 2003). For adolescents, education affects their lives and their communities in many aspects, such as enhanced economic development, increased productivity, lower mortality, and social stability (Little and Green 2009).

### 1.1.4 Behavioral Factors: Physical Activity, Substance Use, and Sexual Practice

Physical activity is associated with improving overall health and lowering risks for chronic health outcomes. Physical activity can improve cognitive function, confidence, and self-esteem. Despite these benefits, lack of sufficient physical activity is one of the main risk factors for death around the world (Byrne et al. 2019). Physical activity is one of the crucial behavioral factors that affect adolescents' health. According to Von Rueden et al. (2006), physical well-being is strongly associated with increased risk of low health-related quality of life. Good physical activity could promote numerous health benefits for children, adolescents, and adults. According to the Centers for Disease Control and Prevention (2020), the benefits of physical activity for adolescents include building strong muscles and bones, reduction of symptoms of depression and anxiety, weight control, cardiorespiratory improvements, and reduction in risks of developing heart disease, type 2 diabetes, cancer, high blood pressure, obesity, and osteoporosis. Guidelines state that adolescents have an excess of 60 min of moderate-to-vigorous intensity exercise daily (Ridley et al. 2008). Adolescents residing in the MENA region are at risk of an increasingly sedentary lifestyle, which leads to a rise in the rates of obesity and other non-communicable diseases.

Substance abuse is a relatively common determinant of adolescent health worldwide. Alcohol, tobacco, and drug abuse and addiction are factors that detrimentally affect adolescent health. Some studies show that young people with relatively higher socioeconomic status are more likely to take up smoking (Hanson and Chen 2007). However, country laws and cultural norms influence access to substances, and thus effective programs and policies are important relevant factors (Viner et al. 2012). Underage drinking is a common worldwide problem and is a dangerous behavior for adolescent health and entire communities. A study conducted in the United States among high school students showed that 29% of students drank alcohol within a period of 30 days and 5% drove after drinking (CDC 2020). Some of the adverse effects of alcohol consumption for youth are disruption of normal growth, alcohol poisoning, memory problems, injuries, and death (CDC 2020; WHO 2020).

The onset of adolescence brings hormonal changes, which influence sexual activity. The ability to remain free from sexually transmitted infections, unwanted pregnancy, and unsafe abortions are all aspects that underline the importance of adolescent sexual and reproductive health (ASRH) (Bearinger et al. 2007). In developed countries, adolescents are educated in the school system about safe sex practices, consent, and sexually transmitted infections (STIs). However, there is a lack of sexual reproductive health education in the MENA region as sexual practices outside of marriage are unconventional and considered a taboo (Bearinger et al. 2007). Adolescents who become sexually active without access to essential health information may risk contracting STIs (Bearinger et al. 2007). Therefore, countries in the MENA region should be encouraged to adopt a Western approach to adolescent sexual and reproductive health.

## 1.2 Adolescent Health and Health-Related Measures

Measuring the health status and health determinants of an individual provide insight into an individual's quality of life. Several instruments provide a quantitative value of health and health risks, which reflect an overall judgment on an individual's health. For example, a public health issue that affects youth worldwide is obesity, as it likely extends into adolescence and adulthood. The body mass index (BMI) is the primary tool used worldwide to assess obesity (Byrne et al. 2019). BMI is the ratio of an individual's weight (kg) divided by their height squared ($m^2$); a BMI ratio of more than 25 indicates overweight, while exceeding 30 defines obesity (CDC 2020). BMI is a good indicator for the measurement of over-nutrition and malnutrition, as well as physical inactivity (Byrne et al. 2019). Physical inactivity is an important risk factor for chronic diseases and obesity (Wolfe 2015). The main objective of measuring physical activity is to estimate energy expenditure, using information on exercise intensity, duration, and frequency. For children and adolescents between the ages of 5 and 18 years, an excess of 60 min daily of moderate-to-vigorous intensity physical activity (MVPA) is recommended (Ridley et al. 2008). Aerobic fitness can be measured by an endurance test (AIHW 2007). Owing to convenience, questionnaires are the dominant method of assessing physical activity despite the limitations, such as recall bias and social desirability bias (Wolfe 2015). Some studies in the MENA region assign a metabolic equivalent value (MET), which allows for accurate measurement of an individual's energy intake (EI) and total energy expenditure (EE) (Ridley et al. 2008). This system allows for accurate measurement of the energy cost associated with distinct physical activities. For example, jogging, running, weight training, and cycling are all examples of vigorous-intensity physical activities (Ridley et al. 2008). Other modifiable and non-modifiable factors influence the health status of adolescents, such as family income, social segregation, and environmental factors (Wagenaar 2003). For example, some families can only afford to purchase calorie-dense unhealthy foods; therefore, they are more susceptible to being overweight and having a relatively high BMI (AIHW 2007).

Another indicator of an individual's health status is oral health as it reflects several factors, including diet, substance use, good hygiene habits, and smoking. The burden of dental caries is on the rise globally, with middle-income countries experiencing the highest prevalence among their adolescent population (Lattanzi et al. 2020). The dental needs of adolescents are overlooked in the MENA region with three out of ten adolescents experiencing dental pain at some point. Most countries in the MENA region lack data on the trends of dental caries; therefore, it is difficult to assess the burden in the region as a whole. Kuwait reported a 14% prevalence of dental caries among their adolescent population, while Riyadh (KSA) reported a prevalence rate of 64.98% (Al-Rafee et al. 2019). Countries in the MENA region need to develop a surveillance program to track the oral health of adolescents. Given that the prevalence of chronic respiratory diseases (CRDs) is on the rise globally, respiratory health is another important measure of adolescent health. Pulmonary disorders that may become prevalent in adolescence include respiratory allergies, asthma, pulmonary hypertension, chronic obstructive pulmonary disease (COPD), and hay fever. While the prevalence of asthma has risen globally, there are variations in the severity and mortality rates between developed and developing nations.

## 1.3 Overview of Upcoming Chapters

Nutrition plays a significant role in adolescents' present and potential well-being as emphasized in this chapter. Fawaz presents information and establishes data using evidence on adolescent nutrition in the MENA region. The author discusses the prevalence of malnutrition and over-nutrition and their link to socioeconomic conditions and highlights the health issues that could result from overweight and obesity, such as hypertension, diabetes, cardiovascular illness, and musculoskeletal conditions. Fawaz points out that the rates of obesity or overweight in teenagers reach 50% in Gulf countries, while countries such as Morocco, Sudan, and Yemen have lower rates (under 16%).

In the following chapter, Abou El Fadl and Fathi Mousa describe the challenges of adolescents' access to oral care and the key determinants that influence their oral health outcomes. The authors point that related social and economic determinants are modifiable by laws and policies and include socioeconomic status, tobacco usage, alcohol consumption, and substance abuse. Adolescents' oral care-seeking behavior and the quality of service they receive are impacted by household income and parental perceptions of the importance of oral health. Those whose mothers have a low level of education are particularly prone to have tooth decay, as they are less likely to adhere to routine oral hygiene practices. The authors report that about 4.6 million (3.8 million are males; 790 thousand are females) adolescents in the MENA region smoke. In Lebanon, 72% of adolescents between the ages of 13 and 17 reported exposure to secondhand smoking; the rates were lower in other nations such as Kuwait (64.3%), the UAE (39.6), Morocco (39.6), and Oman (25.8). The authors also discuss the importance of an oral health surveillance plan since it allows for the assessment of the oral health needs of adolescents. The authors recommend regular surveillance plans in order to capture information on adolescents' pattern of oral care.

A literature review on physical activity among adolescents in the MENA found a high prevalence of physical inactivity (Barakat et al.). Specifically, the authors report that Lebanon and Egypt represent the lowest (65%) and highest prevalence (91%) of physical inactivity in the region, while only 16% of children in the UAE achieved the recommended moderate-to-vigorous physical activity level. Studies from the UAE, Palestine, Morocco, Lebanon, Israel, Kuwait, and KSA suggest significant gender differences. Girls consistently demonstrate lower physical activity levels than boys. The authors report that a common trend that has been normalized among girls across the MENA is adopting a sedentary lifestyle, which may deprive adolescent girls of the opportunities associated with leisure activity and exposure to different forms of outdoor exercise with their cohort counterparts. A study from the UAE suggested that for adolescent girls (11–16 years), low levels of physical activity might relate to various factors, including cultural and weather restrictions and social change of the community in the UAE. The authors recommend outdoor activities to help meet the recommended physical activity levels. Social support has an important influence on adolescent participation in physical activity. Adolescents were more likely to be active when they had the support of their friends and peers.

Literature highlights high rates of overweight among adolescents (15–18 years) in Algeria, Jordan, Kuwait, Libya, Palestine, Syria, and the UAE. These rates range from 9.3% in Algeria to 25.6% in Kuwait. Due to recent urbanization and socioeconomic development, the rates of obesity are on the rise in the MENA region (Khalaf et al.). The Gulf Cooperation Council (GCC) countries (Bahrain, Qatar, Kuwait, Oman, KSA, UAE) have some of the highest prevalence rates or (levels) of overweight and obesity in the world among adolescents. The authors report that the UAE ranks fifth in the world in the obesity prevalence rate of 36%, with an economic burden of $6 billion per year in associated disease cost. Emirati children are almost two times more obese than their American counterparts. The authors report that risk factors for adolescent obesity include ethnicity, genetics, hormonal and metabolic disorders, sleeping disorders, as well as environment. Furthermore, the authors highlight the importance of an effective predictive model that utilizes several algorithms, which aim to predict the prevalence of overweight and obesity rates among adolescents. The classification tree approach can predict overweight or obesity at the age of 14 from BMI calculated at 6 years, as well as the education level and obesity of the mother. The authors state that the BMI value at the age of 6 years was an important predictor in this model.

In the following chapter, Oraby discusses the methods in which adolescents learn about sexual and reproductive health (SRH), patterns and potential consequences of adolescent sexual behavior in the MENA region, and how Islam can help tackle these shortcomings. Parents, schools, and the Internet are the primary channels through which adolescents learn about SRH. Oraby elaborates on a national survey of young people and their parents in Egypt, which revealed that only 7% of boys reported learning about puberty from their fathers, even though 42% of fathers reported talking to their adolescent sons about pubertal changes. An updated survey from 2010 with a sample of 15,000 participants between the ages of 10 and 29 years reported a lack of discussion on sexual and reproductive health from parents. Family members are the primary source of information on puberty for only 5.8% of males. The author states that cultural values, primarily Muslim culture, clash with Western social norms about sex and gender. Many individuals and policy-makers fear that Western social norms could lead to corruption and premature sexualization. Egyptian classrooms were reported as a source of information about puberty for 4.1% of females and 11.5% of males. Oraby states that SRH is often not covered adequately because teachers consider it a taboo topic. The Internet often provides inaccurate information about SRH, with high exposure to online pornography. The author reports that 74% of Egyptian youth between the age of 13 and 24 have watched pornography. Providing adolescents with relevant SRH information is likely to be a difficult process due to the perceived encroachment of Western values. The author recommends a modified Western approach to ASRH that is tailored to the different cultures in the MENA region.

Assessing the growth of adolescents provides an accurate indication of their nutritional status and general health, as highlighted by Cheikh Ismail and Al Dhaheri. The use of anthropometry allows for accurate monitoring of adolescent physical growth while evaluating potential deviations from normal growth such as stunting, underweight, overweight, and wasting. Anthropometric measurements include weight, height, and an individual's BMI. According to the authors, little data is available on adolescent anthropometry and growth in MENA, even though a large body of studies on childhood and adulthood exists. A study in KSA found an increase in the prevalence of stunting among adolescents (5–17 years). A similar study found a low prevalence of stunting in Libyan children (9–11 years). Another study found that one in five children (6–16 years) in Yemen is overweight and obese. In Lebanon, lifestyle and socioeconomic status correlate to obesity and overweight in children and adolescents (6–19 years). The authors report that one of the most challenging health issues of the century is the increased prevalence of overweight and obesity; therefore, it is essential to implement a comprehensive monitoring strategy. It is also important to encourage better dietary habits to reduce the prevalence of wasting and stunting in the MENA region.

In chapter "Oral Health of Adolescents in the MENA Region", Abou El Fadl and Fathi Mousa discuss the distribution and prevalence of common oral issues such as fluorosis, dental caries, periodontal diseases, dental traumatic injuries, and soft tissue lesions among adolescents in the MENA. The authors report that the prevalence of unmet dental needs among adolescents is 84.2% in the Eastern Mediterranean region and 78% in Africa. One in three individuals between the ages of 13 and 18 in MENA experiences dental pain at some point. Although data on dental caries is limited in the MENA, a few researchers have investigated this important topic. A study in Kuwait reported that 14% of 12-year-old children have caries, while a national survey in Egypt reported that around 70% of children between the ages of 4 and 18 experienced some form of tooth decay. The rates of dental caries among adolescents can sometimes vary from province to province; for example, in Jazan (KSA), the prevalence was over 90%, while Riyadh had a prevalence of 64.98%. According to the authors, data on periodontal diseases is limited to a few nations in the MENA region. They report that in Kuwait, 42.5% of 12-year-old children had gingivitis, while the prevalence of the disease reached 65% among high schoolers in KSA. Data on orofacial traumatic injuries is extremely limited; however, one study in Jordan revealed that 14.6% of 12-year-old children suffered from some form of dental trauma. Some countries in the MENA region have excessive fluoride content in water, which is a major cause of dental fluorosis. The authors report that 30.8% of 14-year-old children in Yemen suffered from dental fluorosis due to high levels of fluoride in drinking water (exceeding 10 mg/L). The authors recommend that all nations in the MENA region should develop a surveillance tool for common health determinants that influence oral health.

Chapter 9 describes a cross-sectional study that examines the distribution and prevalence of respiratory and allergic conditions among youths residing in the UAE. The study collected data on the allergy and respiratory profile of 6363 participants between the ages of 13 and 20 attending private and public schools. The prevalence of asthma was 12%, with 6% of participants reporting hospitalization due to asthma. Results demonstrated that the prevalence of rhinitis was 33.2%, rhinitis in the last 12 months was 26%, itchy rash in the last 6 months was 8.4%, and itchy rash in the past 12 months was 7.4%, respectively. The authors state that the increased incidence of allergic disorders and asthma might contribute to a high risk of morbidity and mortality and a decrease in quality of life. Genetics play a role, but the environmental conditions in the region significantly increase the risk of respiratory conditions. Risk factors of respiratory conditions include environmental exposures from crude oil and natural gas production, which is common in the UAE. The authors conclude that self-reported asthma symptoms, wheezing, and other allergy-related diseases are common among adolescents in the UAE mainly due to indoor and outdoor air quality and behavioral practices.

In chapter "The Epidemiology of Diabetes Mellitus Among in Adolescents and Their Determinants in from the Middle East and North Africa Region", Yousufzai and Barakat discuss the prevalence, characteristics, and predictors of type 1 diabetes (T1D) and type 2 diabetes (T2D) among adolescents in the MENA region. T1D occurs due to an autoimmune response that destroys an individual's insulin-secreting β-cells. T2D is more preventable as the progressive resistance to insulin produced within the body characterizes it. The authors point that T2D relates to physical inactivity and nutritional habits. Kuwait's prevalence of T2D among adolescents is 34.9 per 100,000, which exceeds the prevalence reported in Canada and the United States. Among individuals between the ages of 0 and 14 years, KSA, Kuwait, and

Qatar rank among the top ten countries in the world for incidence rates of T1D. Each year 20,800 children and adolescents are diagnosed with T1D in the MENA region, with 150,000 currently diagnosed. The authors report that the prevalence of T1D in KSA is 109.5 per 100,000 for individuals between the ages of 5 and 18. One study estimates that the incidence rate of T1D in Qatar is 23.15 per 100,000 (between 2006 and 2011), while the overall incidence of T2D is 2.9 per 100,000 per year. A study conducted in Egypt estimates the prevalence of T2D among adolescents to be 13.3%. The authors point that risk factors for T1D and T2D include obesity, family history of diabetes, sex, pubertal onset, and vitamin D deficiency. Yousufzai and Barakat conclude that the development of prevention strategies that target modifiable risk factors of diabetes is essential to reduce the mortality and morbidity related to diabetes.

In the last chapter, Maatouk and Assi discuss the challenges that governments face when tackling the rise of HIV and other STIs among key and vulnerable populations in the MENA region. The authors report that of the 240,000 individuals living with HIV in the MENA region, there are approximately 6300 (2.6%) adolescents. The authors recommend the utilization of an HIV self-test (HIVST), which is an intervention strategy that allows patients to manage their health through self-medication or providing care to dependent persons. Vulnerable populations in the MENA region, such as men who have sex with men (MSM) and refugees, often avoid the healthcare system due to stigma. Therefore, HIVST focuses on providing a safe environment for HIV testing. Other programs such as partner notification (PN) services are essential and should be integrated with a multidimensional support system from religious, health, psychological, and legal perspectives.

## 1.4 Putting It in Perspective and Moving Forward

While this book examines adolescent determinants of health and adolescent health status in MENA, topics were limited and failed to encompass the wide range of adolescent health issues that are prevalent in MENA. For example, topics related to the epidemiology of adolescent intentional and unintentional injuries across MENA are missing. There are various reasons these topics are not addressed. In the case of injuries, suicide or intentional injuries are considered taboos due to cultural, religious, and social beliefs. Other reasons that prevent pertinent adolescent health issues from being brought forward to the forefront include limited research and resources, as well as a focus on immediate health-related issues such as the provision of general health assessments, healthcare needs, clean water, and housing for displaced populations, many of whom are adolescents in the MENA region. For example, intentional injuries, violence, and motor vehicle crashes are major contributors to adolescent morbidity and mortality in MENA; however, these issues are not addressed comprehensively with targeted monitoring and surveillance.

Research and policy need to target many areas related to adolescent health in MENA. The *2012 Lancet Adolescent Health Series* can assist in framing the research needs by focusing on key causes of mortality and disability in ages 10–24 years, health-related behaviors and conditions that arise in adolescence, risk and protective environmental factors, indicators of social role shifts related to different forms of health risk, and health policy interventions. Continuous data collection, analysis, and interpretation of major health-related events that affect adolescents, as well as determinants and causes of health-related events, can help policy-makers and public health professionals take action to reduce morbidity and mortality.

The health impacts of physical environments are significant in MENA given that many countries face political unrest and conflicts, which are marked by devastating destructions to housing and water infrastructure systems, as well as major impacts to air quality. One needs to only think back to the 2020 Beirut explosion caused by over 2500 tons of ammonium nitrate, which led to severe destruction of neighborhoods and devastating pollution levels in the surrounding air, soil, and water (Rehman et al. 2021). While similar events have been chronic for decades and have surely impacted the health of adolescents, they are compounded by factors related to social environments, which play significant roles in adolescent health in MENA. For example, low economic prosperity accompanied with high rates of unemployment negatively reflects on employment or skill development opportunities that are open to MENA adolescents. In fact, most MENA societies are built around cultural norms that emphasize education for adolescents and underestimate the importance of summer or part-time jobs or societal volunteer opportunities. Adolescent employment or skill development opportunities are essential in the growth of adolescents and enhance their potential to become productive members of society. In fact, these opportunities are beneficial in allowing adolescents to have alternative role models, occupational values, motivation, and a future career path. Future research into societal determinants of health is an area that has not seen much development in MENA and is crucial in advancing the health of MENA adolescents.

## References

Akseer, N., Al-Gashm, S., Mehta, S., Mokdad, A., & Bhutta, Z. A. (2017). Global and regional trends in the nutritional status of young people: a critical and neglected age group. *Annals of the New York Academy of Sciences, 1393*(1), 3–20.

AlBuhairan, F. S. (2015). Adolescent and young adult health in the Arab region: where we are and what we must do. *Journal of Adolescent Health, 57*(3), 249–251.

AIHW (Australian Institute of Health and Welfare). 2007. Young Australians: Their Health and Wellbeing. Canberra: AIHW.

Al-Rafee, M.A., Al Shammery, A.R., Al Rumikan, A.S., Pani, S.C., 2019. A comparison of dental caries in urban and rural children of the Riyadh region of Saudi Arabia. Front. Public Heal.

Anteghini, M., Fonseca, H., Ireland, M., & Blum, R. W. (2001). Health risk behaviors and associated risk and protective factors among Brazilian adolescents in Santos, Brazil. *Journal of adolescent health, 28*(4), 295–302.

Bchir A, Bhutta Z, Binka F, Black R, Bradshaw D., and others. 2006. "Better Health Statistics Are Possible." *The Lancet* 367 (9506): 190–3.

Bearinger, L., Sieving, R., Ferguson, J., & Sharma, V. (2007). Adolescent health 2 global perspectives on the sexual and reproductive health of adolescents : Patterns, prevention, and potential. *The Lancet* (British Edition), 369(9568), 1220–1231.

Bhutta, Z., Victora, C., Boerma, T., Kruk, M., Patton, G., Black, M., Sawyer, S., Horton, S., Black, R., & Horton, R. (2019). Optimising the continuum of child and adolescent health and development. *The Lancet* (British Edition), 393(10176), 1080–1082. https://doi.org/10.1016/S0140-6736(19)30488-X

Blum, R. W., Halcón, L., Beuhring, T., Pate, E., Campell-Forrester, S., & Venema, A. (2003). Adolescent health in the Caribbean: risk and protective factors. *American Journal of Public Health*, *93*(3), 456–460.

Boerma J T, Stansfield S K. 2007. "Health Statistics Now: Are We Making the Right Investments?" *The Lancet* 369 (9563): 779–86.

Bronfenbrenner U., 1979. The Ecology of Human Development: Experiments by Nature and Design. Cambridge, MA: Harvard University Press.

Bruneforth, M., & Wallet, P. (2010). Out-of-school adolescents. *Montreal: UNESCO Institute for Statistics.*

Byrne, R., Bell, L., Taylor, R., Mauch, C., Mihrshahi, S., Zarnowiecki, D., Hesketh, K., Wen, L., Trost, S., & Golley, R. (2019). Brief tools to measure obesity-related behaviours in children under 5 years of age: A systematic review. *Obesity Reviews*, 20(3), 432–447. https://doi.org/10.1111/obr.12801

Cameron, L., & Williams, J. (2009). Is the Relationship between Socioeconomic Status and Health Stronger for Older Children in Developing Countries? Demography, 46(2), 303–324. https://doi.org/10.1353/dem.0.0054

Centers for Disease Control and Prevention. (2020). *Physical Activity Facts*. Retrieved from CDC website: https://www.cdc.gov/healthyschools/physicalactivity/facts.htm

Christian, P., & Smith, E. R. (2018). Adolescent undernutrition: global burden, physiology, and nutritional risks. *Annals of Nutrition and Metabolism*, *72*(4), 316–328.

Currie, C., Molcho, M., Boyce, W., Holstein, B., Torsheim, T., & Richter, M. (2008). Researching health inequalities in adolescents: The development of the Health Behaviour in School-Aged Children (HBSC) Family Affluence Scale. *Social Science & Medicine* (1982), 66(6), 1429–1436. https://doi.org/10.1016/j.socscimed.2007.11.024

Gakidou, E., Cowling, K., Lozano, R., & Murray, C. J. (2010). Increased educational attainment and its effect on child mortality in 175 countries between 1970 and 2009: a systematic analysis. *The lancet*, *376*(9745), 959–974.

Galler, J. R., Koethe, J. R., & Yolken, R. H. (2017). Neurodevelopment: the impact of nutrition and inflammation during adolescence in low-resource settings. *Pediatrics*, *139*(Supplement 1), S72–S84.

Hanson, M. D., & Chen, E. (2007). Socioeconomic status and health behaviors in adolescence: a review of the literature. *Journal of behavioral medicine*, *30*(3), 263.

Hawkins J D, Oesterle S, Brown E C, Arthur M W, Abbott R D., and others. 2009. "Results of a Type 2 Translational Research Trial to Prevent Adolescent Drug Use and Delinquency: A Test of Communities That Care." *Archives of Pediatrics and Adolescent Medicine* 163 (9): 789–98.

Kravdal, Ø. (2002). Education and fertility in sub-Saharan Africa: Individual and community effects. *Demography*, *39*(2), 233–250.

Little, A. W., & Green, A. (2009). Successful globalisation, education and sustainable development. *International Journal of Educational Development*, *29*(2), 166–174.

Lattanzi, A., Silveira, F., Guimarães, L., Antunes, L., Santos Antunes, L., & Assaf, A. (2020). Effects of oral health promotion programmes on adolescents' oral health-related quality of life: A systematic review. *International Journal of Dental Hygiene*, 18(3), 228–237. https://doi.org/10.1111/idh.12440

Murray C J L. 2007. "Towards Good Practice for Health Statistics: Lessons from the Millennium Development Goal Health Indicators." *The Lancet* 369 (9564): 862–73.

Murray, L., & Lopez, AD. (2013). Measuring the Global Burden of Disease. *The New England Journal of Medicine* 369: 448–57.

Najman, J. M., Aird, R., Bor, W., O'Callaghan, M., Williams, G. M., & Shuttlewood, G. J. (2004). The generational transmission of socioeconomic inequalities in child cognitive development and emotional health. *Social science & medicine (1982)*, *58*(6), 1147–1158. https://doi.org/10.1016/s0277-9536(03)00286-7

Pal, A., Pari, A. K., Sinha, A., & Dhara, P. C. (2017). Prevalence of undernutrition and associated factors: A cross-sectional study among rural adolescents in West Bengal, India. *International Journal of Pediatrics and Adolescent Medicine*, *4*(1), 9–18.

Patel, V., Flisher, A. J., Nikapota, A., & Malhotra, S. (2008). Promoting child and adolescent mental health in low and middle income countries. *Journal of child psychology and psychiatry*, *49*(3), 313–334.

Patton G C, Coffey C, Carlin J B, Sawyer S M, Williams J., and others. (2011). "Overweight and Obesity between adolescence and young adulthood: a 10-year prospective cohort study".

Rehman A., Ahmed, R., Ma, K., Xu, S., Aslam, M., Bi, H., Liu, J., Wang, J. (2021). Ammonium nitrate is a risk for environment: A case study of Beirut (Lebanon) chemical explosion and the effects on environment. *Ecotoxicology and Environmental Safety*, 210 (1), 111834.

Ridley K., Ainsworth, B., Olds, T. (2008). Development of a Compendium of Energy Expenditures for Youth. *International Journal of Behavioral Nutrition and Physical Activity*, 5: 45.

Shope J T, Molnar L J, Elliott M R, Waller P F. 2001. "Graduated Driver Licensing in Michigan." *Journal of the American Medical Association* 286 (13): 1593.

Starfield, B., Riley, A. W., Witt, W. P., & Robertson, J. (2002). Social class gradients in health during adolescence. *Journal of Epidemiology & Community Health*, *56*(5), 354–361.

Rosen J E, Levine R. 2010. Mainstreaming Adolescent Girls into Indicators of Health Systems Strengthening. Washington, DC: Center for Global Development

Viner, R. M., Ozer, E. M., Denny, S., Marmot, M., Resnick, M., Fatusi, A., & Currie, C. (2012). Adolescence and the social determinants of health. *The lancet*, *379*(9826), 1641–1652.

Von Rueden, U., Gosch, A., Rajmil, L., Bisegger, C., & Ravens-Sieberer, U. (2006). Socioeconomic determinants of health related quality of life in childhood and adolescence: results from a European study. *Journal of Epidemiology & Community Health*, *60*(2), 130–135.

Wagenaar A C. 2003. "Research Affects Public Policy: The Case of the Legal Drinking Age in the US." *Addiction* 88 (s1): 75S–81S.

Wolfe, DJ (2015). The Effects of Socioeconomic Status on Child and Adolescent Physical Health: An Organization and Systematic Comparison of Measures. *Social Indicators Research*, 123(1), 39–58. https://doi.org/10.1007/s11205-014-0733-4

World Health Organization. (2020). *Overweight and obesity*. Available at: https://www.who.int/gho/ncd/risk_factors/overweight_adolescents_text/en/

# Part II
# Determinants of Adolescent Health in the MENA

# Chapter 2
# Adolescent Nutrition: A Focus on MENA

**Mirna Fawaz**

## 2.1 Introduction

While adolescence is generally deemed a healthier age, it is projected that more than 1.4 million adolescents die annually (97% of such fatalities take place in low- and middle-income states), and a much larger amount are suffering from diseases. Adolescence correlates with dramatic shifts that influence adult health outcomes. Such changes originate from factors that evolve through adolescent years (Obermeyer 2015).

Adolescence is a period of freshly found individuality and freedom of expression. This places adolescents in a category that is vulnerable to extrinsic pressures, especially from the media, school, and their contemporaries Adolescents undergo cycles of accelerated development correlated with shifts in hormones, behaviors, and emotions. Healthy eating behaviors throughout adolescence are an expansion of eating patterns that arise during early development, and thus nutrition could have a vital role in acute or chronic disease incidence, development, and even prognosis. For instance, malnutrition affects the prevalence of both communicable and non-communicable disorders and death rates and influences the growth of individuals and communities (Szabo et al. 2019).

Obesity and overweight are increasingly becoming serious concerns among adolescents in several regions of the Arab world, arising at times in the very same state parallel to under-nutrition within various economic classes. The Middle East and North Africa (MENA) region comprises of low- and middle-income states that face

M. Fawaz (✉)
Nursing Department, Faculty of Health Sciences, Beirut Arab University, Beirut, Lebanon
e-mail: mirna.fawaz@bau.edu.lb



C. Barakat et al. (eds.), *Adolescent Health in the Middle East and North Africa*, Global Perspectives on Health Geography,
https://doi.org/10.1007/978-3-030-92107-1_2

a combined pressure of malnutrition (i.e., stunting, excess, and micronutrient inadequacies) and over-nutrition (i.e., overweight and obesity) and related chronic diseases. The nutritional condition of adolescents around MENA demonstrates inequities in the nutritional status of adolescents between and within nations. It is notable that the greatest rise in obesity prevalence was found in low- and medium-income states, with the MENA being one of the areas with the largest occurrence of overweight among adolescents. It should be recognized that while part of the growing rate of adolescent obesity in more deprived than richer societies may be due to globalization and epidemiological change, it might also be due to the issue of stunting among adolescents in the developing states. Disparities in the incidence of obesity in states of the Middle East and North Africa (MENA) region represent the various stages of the dietary transition – the mechanism by which economic reality, such as labor industrialization and improved food supply, contributes to the abolition of conventional diets, elevated sugar and fat intake, diminished cereal and fruit intake, and physical engagement (Farrag et al. 2017). In addition, the Middle East and North Africa (MENA) has been inundated with unresolved wars for several years, further influencing the nutritional status of the most marginalized populations. On the other hand, the combined pressure of nutritional disorders and the chronic issue of micronutrient deficiency should be recognized as factors for increased body mass index (BMI) in adolescents in developed countries of the MENA region (UNICEF 2020). Therefore, nutrition has a significant influence on adolescents' present and potential well-being. A good nutritious lifestyle with balanced eating habits throughout adolescent years has the ability to reduce any nutrient deficiencies and linear growth deterioration created within the first years of adulthood, thus restricting unhealthy activities that lead to the incidence of consequential chronic illnesses. Prioritizing youth well-being is an economic investment: enhancing the health security for adolescents today, enhancing well-being and prosperity in their potential adult years, and diminishing health threats to their offsprings. Yet, various factors might influence the nutritional intake and therefore overall health of adolescents. This chapter presents information and establishes data from evidence on adolescent nutrition in the MENA region. In doing so, an introduction into the general topic of nutrition and adolescence is initially presented, followed by our research methodology and findings.

## 2.2 Nutrition and Adolescence

### *2.2.1 Factors Influencing Eating Behaviors of Adolescents*

Eating patterns and dietary choices that affect the caloric intake and the consumption of nutrients are normally determined over a span of years and especially during adolescence. There are several influences that affect food preferences and dietary consumption in this age category, like general awareness of nutritional intake, social class, urban or rural dwelling, family structure, ethnic and traditional activities, and engagement in recreation. Living environments during adolescence is recognized to have a major impact on nutritional preference, where families act as role models, which enhances and promotes healthy eating behavior development. Adolescent food preferences are also influenced by peer pressure and heightened social interaction. Anorexia or refusal to eat meals, bingeing, eating anything accessible, and consuming comfort or junk foods are among the food-related reactions to adolescent lifestyle stresses. However, adolescents are generally in command of their food patterns relative to younger individuals, as they explore more in food preferences and detract from the norm of consuming three meals a day. Thus, food patterns impair the consumption of nutrients because the source of food differs; thus, nutritional requirements shall be safeguarded (Ferris et al. 2017).

### *2.2.2 Adolescent Nutritional Requirements*

During adolescence, the developmental peak involves accelerated growth of tissue with specific nutritional requirements. The consumption of energy must adhere to the requirements of adolescents especially whether they participate in physical activity or regular exercise. In general, adolescent males have a higher calorie intake than adolescent females, leading to greater changes in height, weight, and lean body mass. The US Departments of Agriculture/Health and Human Services (USDA/HHS) Dietary Guidelines of 2010 find that 50% or more of the overall dietary calories must originate from carbohydrates, with no more than 10–25% of the energy extracted from sugars, such as sucrose and high-fructose corn syrup. Protein demands per unit of height are maximum for females in the age bracket of 11–14 years and for males in the age group of 15–18 years, which leads to the normal peak height velocity. Population-sensitive diets, such as dietary reference diets (DRIs) in the Americas, do not identify clear specifications for overall fat consumption but provide guidelines for the consumption of linoleic (n-6) and-linolenic (n-3) polyunsaturated fatty acids (López-Sobaler et al. 2019). The US DRI for calcium is 1300 mg/day for 9–18-year-olds, and the recommended dietary allowance (RDA) for iron is 8 mg/day for 9–13-year-olds, 11 mg/day for 14–18-year-old males, and 15 mg/day for 14–18-year-old females, because the advent of menstrual cycle places increased iron requirements on women. The US zinc RDA for males and females aged 9–13 years is approximately 8 mg/day, whereas the RDAs are 11 and 9 mg/day for males and females aged 14–18 years, respectively. Males and females aged 9–13 years must obtain 600 g/day to maintain optimal body uses of vitamin A, females aged 14–18 must obtain 700 g/day, and males aged 14–18 must absorb 900 g/day. The recommended daily allowance for vitamin E for 9–13-year-olds is 11 mg/day for the United States and 15 mg/day for 14–18-year-olds. US RDA for vitamin C is 45 mg/day for 9–13-year-olds, 75 mg/day for 14–18-year-old males, and 65 mg/day for 14–18-year-old females. The US folate RDA is 300 g/day for ages 9–13 and 400 g/day for ages 14–18. All macro- and micronutrients are essential to maintain the normal functions of the body as well as promote proper growth rates, where any imbalances can lead to aliments, some of which might develop into chronic concerns (Kathleen and Raymond 2016).

## Macronutrients

### Carbohydrates

Carbohydrates comprise the primary fuel supply for adolescents and typically represent 55% of the total consumption of calories. The glucose and fructose monosaccharides, found in fruits and vegetables, are sources of "healthy" carbohydrates. Their biochemical impact is distinct: while they emit approximately the same level of energy, glucose generates more insulin and is digested in both tissues, whereas the liver metabolizes fructose. Fructose intake, used in soft beverages, is blamed for adolescent obesity. In most healthy diets, which contain fruits, cereals, and dairy, disaccharides, sucrose, lactose, and maltose are present. The most prevalent polysaccharide is starch, which constructs complex carbs alongside fibers. Carbohydrates function primarily at the core of hypothalamic satiety and influence other food intake afterward (Das et al. 2017).

### Protein

Proteins are important to every individual as they are the basic components of the body's tissues and cells. These comprise essential muscle and other tissue constituents of critical body fluids, such as blood. Protein requirements are significantly correlated with the growth rate during adolescence than with actual age. When the consumption of energy is inadequate for some cause (e.g., nutritional shortages, medical illness, or efforts to reduce weight), nutritional proteins can be needed to satisfy the energy requirements and will not be necessary to synthesize new tissues or to restore tissues. This will result in lower rate of growth and reduction in lean muscle mass given an evident sufficient consumption of proteins. Protein digestion is responsive to calorie reduction in rapidly growing adolescents; for example, during a deficient intake of energy, protein is utilized for energy needs leading to undernutrition. The primary forms of animal and vegetable protein, such as beef, poultry, seafood, dairy, corn, grains and nuts, beans, and cereals, produce 20–25% of the overall energy and should be eaten two to three times daily (Ozdemir 2016).

### Fat and Lipids

Fat is an essential dietary element which performs a variety of body processes. Fat is a condensed energy supply, so it offers more than double the energy provided by either protein or carbohydrate per unit mass. Only a tiny quantity of fat (9 kcal/g of fat), if obtained in a physically fit adolescent's diet, will fulfill the required energy.

Lipids have an important caloric feature performed in oils, soybean oil, butter, margarine, bacon, lard, sausages, creams, sauces, burgers, and mayonnaise by saturated and non-saturated fats (Ozdemir 2016).

### Micronutrients

Minerals

Throughout puberty, all mineral requirements increase. During the years of growth peak, adolescents were shown to ingest higher amounts of calcium, iron, zinc, magnesium, and sodium in comparison with that during other ages. In adolescence, calcium and iron are especially required. Calcium is required for bone production. With a further rise of 2 g/dl in hemoglobin in the male and 1 g/dl in the female, the corresponding growth need alone is 0.7 mg/day for males and 0.45 mg/day for females, although mandatory liabilities rise with age as well. Throughout the phenotype manifestation of the genetic capacity of bone tissue development and calcification, sufficient consumption of these nutrients during the adolescent cycle is important (Gupta et al. 2018).

Vitamins

In adolescence, the vitamin demands are all boosted due to enhanced anabolism and caloric burn. Many causes, such as physical exercise, gestation, oral contraceptives, and serious illnesses, often lead to this rise. Throughout the pubertal peak, vitamin A, B, C, and D demands are high, with cellular differentiation and bone development. Consumption of folic acid is regularly recommended for pregnant adolescents. Thiamine, riboflavin, and niacin are necessary to extract energy from carbohydrates in a growing amount. Folacin and B12 are vital in the synthesis of DNA and RNA. Premenstrual syndrome (PMS) could be minimized if adolescent girls take 100 mg of vitamin B per day. Bone development demands vitamin D, while newly established cells' morphological and chemical integrity is dependent on the supply of vitamins A, C, and E. Such vitamins will, in most instances, be given by a deliberate diet without vitamin supplementation being needed. Vegetables, whole cereals, milk, nuts, poultry, eggs, and grains are primary sources. Water, juices, coconut milk, and other fluid sources should be supplied, averaging 4–6 glasses per day (Hans and Jana 2018).

## 2.3 Nutrition Screening

The Society for Adolescent Health and Medicine suggests regular reviews of wellness screening for all adolescents. Initial checks must involve height, weight, and body mass index measures focused on profiles of population-dependent growth.

Information on portion sizes, weight management habits, and excessive feeding can be elicited using a culturally aware, caring, and non-judgmental method. Daily meal schedules, food preparation techniques, consumer habits, and the dietary impact of society and faith shall be considered focusing on trends such as the use of sweeteners, beverages, and alcohol. Thus, comprehensive assessments can detect certain nutritional defects and chronic disorders among adolescents needing intervention (Kumar et al. 2018).

## 2.4 Nutrition and Eating Disorders in Adolescents

Adequate nutrition is important throughout adolescence, as development and growth perform vital roles in maintaining average adult weight and reproductive capacity during this time. The emergence of an eating disorder, such as anorexia nervosa or bulimia nervosa, is a significant impediment to healthy eating among adolescents. Anorexia nervosa entails extreme underweight, apprehension of weight growth, and poor self-esteem. Likewise, bulimia nervosa involves binge consumption accompanied by inappropriate compensation. Both are best tackled to resolve the medical, psychological, and dietary aspects by an interdisciplinary team of experts. Predispositions for developing adolescent eating disorder include gender, excessive dieting, early puberty, personality, obsessiveness, bullying, poor self-confidence, psychological traumas, and discord in the household (Leme et al. 2018). Other nutritional concerns are prevalent among adolescents and mainly are represented by obesity and undernutrition (Table 2.1).

While family therapy is widely used medically for adolescents with anorexia nervosa, only one type of family therapy has been routinely studied in cohort studies – originally designed by the Maudsley team at the Institute of Psychiatry in London and manually administered as family-based treatment (FBT) (Lock 2015).

**Table 2.1** Difference between anorexia nervosa and bulimia nervosa

| Anorexia nervosa | Bulimia nervosa |
|---|---|
| Lower weight | Normal weight |
| Early presentation | Late presentation |
| Never asks for help | May ask for help |
| Early onset | Late onset |
| Can be premenarchal | |
| Can afflict males | Generally afflict girls |
| Acute or chronic | Varying course |
| No prior ailment | Previous anorexia nervosa |
| Linked with stress, obsessive compulsive disorder, depression | Linked with depression, self-harm, substance misuse |
| Prognosis poor without early treatment | Up to 60% respond to specific treatment |

Nagl et al. (2016)

The purpose of the therapy is to recognize and enable the family to change conditions that might have precipitated the problem. The findings of the strategy indicate that 80–90% of patients with decent to outstanding improvement are successful for adolescents in the short term and that therapy outcomes are sustained on the long run (Jewell et al. 2016). On the other hand, cognitive behavioral therapy (CBT) is proposed to be of value to adolescents who suffer from anorexia nervosa (AN) (Agras 2019). The reason for the method is that the AN patient has skewed perception of form and weight and overestimation of slenderness (Dalle et al. 2020). That, in fact, contributes to excessive dieting and weight loss due to overexercising. CBT for AN tackles these mental and skewed thoughts by standardizing dietary habits, tracking food consumption through food logs, collective problem-solving, behavioral tests, and cognitive transformation. CBT modifications for adolescents include the participation of caregivers, the use of real-time tracking, and exposure to behavioral dilemmas within the extended CBT paradigm. Medication trials for adolescents with AN are minimal, and the findings have not been positive to date (Linardon et al. 2017). While drugs in most psychoactive groups have been tested in small case series or in adult studies, none of them reliably show systemic efficacy. There was optimism for the prospects for SSRIs (e.g., fluoxetine) to be as effective as possible in avoiding weight loss following acute weight maintenance, although a major review now shows that this is unrealistic (Sysko et al. 2019).

As for bulimia nervosa (BN), CBT was also adjusted for adolescents and then manually employed in a case series of adolescents with BN who find abstinence rates (30–40%) close to those in adult trials. Just one RCT evaluated directed self-help CBT to FBT for BN in adolescents and observed comparable end-of-treatment and follow-up effects for both classes, but CBT was more cost-efficient. Overall progress was comparable to that recorded in adults treated with CBT with abstinence at 6-month follow-up to about 30% (Hail and Le Grange 2018). Medications are also prescribed for the treatment of BN. In specific, antidepressants have been studied in a number of BN trials. Antidepressants are beneficial and effective for BN, while doses were typically higher than those commonly recommended for depression, and the result seems to be linked to preventing binge incidents rather than modifying mental state. Just one minor case series illustrated the appropriateness and potential value of antidepressants (fluoxetine) for adolescent BN, although this was in the form of an inpatient interdisciplinary recovery facility (Gorrell and Le Grange 2019).

Obesity is described as an anomalous or disproportionate buildup of fat, which poses a health hazard. The body mass index (BMI) is a significant obesity screening tool. Obesity is a dynamic physical condition that has hereditary and environmental determinants. An elevated calorie consumption (e.g., diets rich in saturated fat and sugar) and inactive lifestyle are extrinsic variables that lead to childhood and adolescent obesity (Farrag et al. 2017).

Undernutrition signifies inadequate energy and food consumption to fulfill the requirements of a person to preserve healthy nutrition. Undernutrition is interchangeably used for malnutrition. Variables correlated with adolescent undernourishment in developed countries are poor socioeconomic circumstances in the family, intermittent food scarcity, child labor (family income-poverty marker), chronic diseases, insufficient understanding of the long-term effects of adolescent undernourishment, the availability and quality of food, and exposure to health and nutrition facilities (Leme et al. 2018).

### 2.4.1 Management of Nutritional Disorders

These nutritional disorders require collaboration from interdisciplinary team members to provide holistic care. Optimally, adolescents with dietary problems should be assigned to an interdisciplinary team specialized in adolescent health, comprising a physician or nurse specialist, a dietician, and a psychiatrist. It is necessary to educate adolescents that there are no "healthy foods" or "poor foods" but rather evaluate their choices beforehand. Providers must recognize that adolescents suffering from food deprivation may not be able to reliably adopt instructions for daily eating habits with all food classes being included. Nevertheless, irregular eating habits attributable to food scarcity can contribute to dysfunctional feeding, either by perpetuating dietary restraint or by promoting binge when there is food. Adolescents and parents should be informed that it is the combination of all food types that includes all the nutritional requirements for well-being and that restrictive diets can lead to micronutrient, macronutrient, or energy deficiencies that can be particularly harmful in development cycles. Therefore, adolescents who obtain deficient amounts by food alone, or who have biological or medical proof of micronutrient insufficiency, should be given micronutrient supplements (Golden et al. 2016; Das et al. 2017).

## 2.5 Methodology

In order to meet our research objectives of presenting information and established data from evidence on adolescent health nutrition in the Middle East and North Africa (MENA) region, a review of the current MENA-based research has been undertaken through accessing PubMed, Google Scholar, and EBSCO using the keywords "nutrition," "teens," "teenagers," "adolescents," and "MENA" and country names and technical terms for conditions such as bulimia and anorexia, where the relevant researches were filtered to turn up in the past 10 years. The research strategy yielded 3780 publications, of which 10 were based in the MENA. Our rather narrow aim was to recognize concerns of specific interest to the MENA region and to demonstrate important topics for prospective study. We also looked for global and international analyses, hand-searched journal articles of key regional studies, and collected data from comparable recent surveys.

## 2.6 Findings and Discussion

Upon review of the relevant literature, this paper has resulted in the acknowledgment of diverse nutritional problems among adolescents in the Middle East and North Africa (MENA) region that have been precipitated by the current socioeconomic conditions and political unrest in the region as well as recent changes in the dietary pattern that have been produced by the globalization in the past decades. This notion calls for various national and regional actions to be taken.

### *2.6.1 Nutrition in the MENA*

According to evidence from the 2010 Global Burden of Disease, transportation deaths, other accidental fatalities, and cardiac/circulatory disorders are the main three reasons of mortality among male adolescents aged 15–19 years in the MENA region (Arthur 2014). Among female adolescents aged 15–19 years, infectious diseases (diarrhea, respiratory diseases, and meningitis) are the major cause of death – part of the unresolved problem of communicable diseases that persist in the area. The second and third factors of mortality are coronary and circulatory disorders and accidental accidents. The proportion of deaths from cardiovascular and circulatory disorders among MENA adolescents (both male and female) is almost three times greater than the worldwide levels. Diabetes and cancer also lead at higher proportions than the world average. Malnutrition and over-nutrition are significant predispositions for cancer, death, good growth, and development (Black et al. 2013). Malnutrition and over-nutrition are popular in the MENA area, even within the same states, and often within the same family. Adolescent malnutrition is common in low-income states; in countries suffering humanitarian disasters such as Iraq, Somalia, and Sudan; and among poor communities in prosperous countries. Obesity and overweight have been major community health issues for MENA adolescents (Mirmiran et al. 2010), leading to elevated levels of early metabolism disorders and elevated blood pressure. As a main reason for cardiovascular/circulatory illnesses, diabetes, and musculoskeletal conditions, the ratio of DALYs among adolescents in these circumstances may be significantly higher and in line with evidence among adults in the area (Mokdad et al. 2014). Global school-based student health surveys (GSHS) have found rates of overweight/obesity among adolescents of almost 50% in Gulf countries; low rates (<16%) in nations such as Morocco, Sudan, and Yemen; and between one-fourth and one-third in others, aligned with research from Bahrain, Jordan, the United Arab Emirates (UAE), and Saudi Arabia. A comprehensive analysis found that the levels of overweight or obesity between children and adolescents in MENA were only surpassed by North America and South Latin America (Fleming et al. 2013). The World Health Organization (WHO) indicated that obesity rates must be treated very seriously in East Mediterranean states, where 18 nations form the vast bulk of the Middle East and North Africa region (MENA) have undergone a dietary change. Various research have been established in the field. For instance, the research by Boodai and Reilly (2013) is the first experiment in Kuwait on obesity management. The therapy was meant to eliminate sedentary habits, raise physical exercise levels, and boost health, where the controls sought attention from primary health care while the experiments engaged in focus groups with an interdisciplinary team. The analysis found that intervention had no major control-related impact on the body mass index (BMI) Z-score, and did not lead to any substantial increases in waist girth and blood pressure. The research emphasized the need for inclusion of obese adolescents and their parents in the programs being tested and the necessity for longitudinal research on obesity. In addition, Kuwait has launched a collaboration among its Medicare system, education institutions, industry, and public sector called the Kuwait-Scotland eHealth Innovation Network (KSeHIN), which aims to provide a bundle of healthcare resources, awareness, and clinical trials, which includes around 300 children annually (Conway et al. 2014). Moreover, Habib-Mourad et al. (2014) performed a clinical experiment in Lebanon to determine the efficacy and usefulness of the Health-E-PALS initiative to encourage balanced food and physical exercise focusing on the principle of social cognition among adolescent students. Participants in the treatment group reported consuming less chips and sugar-sweetened drinks bought and eaten. The procedure had little impact on physical exercise and screen time, and there were no records of any differences in body mass index (BMI) among post-intervention classes (Habib-Mourad et al. 2014). A 3-year action research was carried out in schools of Sousse, Tunisia. The treatment strategy contained both training activities and behavioral improvements. Students were motivated to engage in physical exercise by coordinated games of afterschool football. Nutritious eating patterns have been promoted by providing alternate healthy products in school cafeterias. Post-intervention survey found that the consumption of fruits and vegetables in the treatment group improved substantially, whereas the control group was flat (Maatoug et al. 2015). Likewise, a study in urban sites in seven Arab countries found that the percentages of male adolescents who were overweight or obese were lowest in Algiers City, Algeria (13.4%), and highest in Kuwait City, Kuwait (60.4%), whereas the number of overweight or obese female adolescents varied from 16% in Al-Khalil, Palestine, to 41.4% in Kuwait City, Kuwait (Rey-López et al. 2019).

### 2.6.2 *Future Research Areas*

Adolescents now are more vulnerable than in the past years due to nutritional problems, excessive alcohol usage, communicable diseases, and other threats. There are broad disparities in the incidence of predispositions to chronic ailments in adulthood, alcohol misuse, obesity, and sedentary habits within areas of adolescent health profiles. Future work is needed to better understand the adolescent nutrition needs in varying contexts. Investing in this generation can yield dividends for generations to come. Future studies should identify regionally and culturally relevant interventions for all nutritional disorders in every nation. Researchers should investigate strategies for avoiding pitfalls of weight-based nutritional counseling, which may inadvertently encourage disordered eating. Studies are required to investigate how often social factors of health lead to metabolic problems in adolescence, such as traumatic experiences in the childhood or oxidative stress from socioeconomic injustice. Experts must explore methods to prevent weight-based dietary consultation drawbacks that can unintentionally promote maladaptive feeding.

### 2.6.3 Challenges

The nutritional condition of adolescence around the Middle East and North Africa (MENA) shows disparities within and inside countries in the nutritional health of children under 5. Action reports across 13 countries (Algeria, Djibouti, Egypt, Iran, Iraq, Jordan, Lebanon, Libya, Morocco, Syria, Tunisia, the State of Palestine, Yemen) show that the biggest success has been in eradicating severe deprivation and ensuring access to better hygiene. For several years, the area has been overwhelmed by persistent wars, thus impacting the nutritional condition of the most troubled area. The implications of chronic MENA conflict are not restricted to death and disability but have indirect secondary repercussions. Nutrition systems disperse from conflict and non-conflict states, while the former tend to have initiatives based on urgent hunger care and preserving lives. In addition, these states are subjected to broader subnational disparities, particularly within quartile of wealth and residence area (UNICEF 2020).

### 2.6.4 Recommendations for MENA Adolescent Nutrition Programming

1. Ideally, adolescents must be provided with dietary monitoring by a professional healthcare practitioner annually.
2. Exclusive diets which are not appropriate from a medical, religious, or cultural point of view should be avoided.
3. Schools, public service agencies, legislatures, and NGOs have a vital part in the detection and care of juvenile dietary problems.
4. Healthcare professionals must support increasing exposure to, and coverage of, adolescent nutritional programs in high-income countries (HIC) and low- and middle-income countries (LMIC).
5. Education exercises should be simple and enjoyable and should prove that nutritious meals are inexpensive and simple to cook and can be flavorful.

There is an ample proof of initiatives in the Middle East and North Africa (MENA) area to tackle vigorously the child obesity issue. A clear indication of this initiative is the Kuwait-Scotland e-Health Innovation Network (KSeHIN). A significant public health hurdle, authors say, is how to create positive dietary habits among adolescents. Peer participation, self-assessment, and encouragement as well as environmental strategies complementing behavioral change and community engagement are essential to the effectiveness of adolescent nutrition education. Continued attempts to decrease the incidence of childhood obesity are significant. In order to achieve that, recommendations would include making instructional resources useful but still enjoyable, engaging, and important to the public.

## 2.7 Conclusion

Over the past decade, the MENA area experienced a startling and well-reported increase in the levels of overweight and obese adolescents. Although statistics are not comprehensive, the rate of growth tends to be quicker than that observed among adults in the same area and is relatively fast in certain states, notably in the Gulf area, particularly in Kuwait, Qatar, and Saudi Arabia. The rate of childhood obesity in many MENA nations is observably greater than in Western nations. Economic influences and shifts in nutrition and activity, as well as how children and adolescents expend their time in sedentary habits versus active exercises, seem to be key drivers of this dramatic, recent rise in overweight and obesity levels.

## Reflection Questions

1. What characterizes adolescence as a developmental stage?
2. What is the difference between macro- and micronutrients needed during adolescence?
3. How are eating disorders precipitated during adolescence?
4. How do the cultural and sociopolitical contexts affect nutrition in the MENA region?
5. How can nutrition be inducted on the policymaking table and what actions could be taken?

## References

Agras, W.S., 2019. Cognitive behavior therapy for the eating disorders. *Psychiatric Clinics*, *42*(2), pp.169–179.

Arthur, M., 2014. Institute for health metrics and evaluation. *Nursing Standard (2014+)*, *28*(42), p.32.

Black, R.E., Victora, C.G., Walker, S.P., Bhutta, Z.A., Christian, P., De Onis, M., Ezzati, M., Grantham-McGregor, S., Katz, J., Martorell, R. and Uauy, R., 2013. Maternal and child undernutrition and overweight in low-income and middle-income countries. *The lancet, 382*(9890), pp.427–451.

Boodai, S.A. and Reilly, J.J., 2013. Health related quality of life of obese adolescents in Kuwait. *BMC pediatrics, 13*(1), p.105.

Conway, N.T., Al Wotayan, R., Alkuzam, A., Al-Refaei, F.F., Badawi, D., Barake, R., Bell, A., Boyle, G., Chisholm, S., Connell, J. and Emslie-Smith, A., 2014. The Kuwait-Scotland eHealth Innovation Network (KSeHIN): a sustainable approach to quality improvement in healthcare. *Quality in primary care, 22*(1), pp.43–51.

Dalle Grave, R., Conti, M. and Calugi, S., 2020. Effectiveness of intensive cognitive behavioral therapy in adolescents and adults with anorexia nervosa. *International Journal of Eating Disorders, 53*(9), pp.1428–1438.

Das, J.K., Salam, R.A., Thornburg, K.L., Prentice, A.M., Campisi, S., Lassi, Z.S., Koletzko, B. and Bhutta, Z.A., 2017. Nutrition in adolescents: physiology, metabolism, and nutritional needs. *Annals of the New York Academy of Sciences, 1393*(1), pp.21–33.

Farrag, N.S., Cheskin, L.J. and Farag, M.K., 2017. A systematic review of childhood obesity in the Middle East and North Africa (MENA) region: Prevalence and risk factors meta-analysis. *Advances in pediatric research, 4.*

Ferris, K.A., Babskie, E. and Metzger, A., 2017. Associations between food-related parenting behaviors and adolescents' engagement in unhealthy eating behaviors: The role of nutrition knowledge. *The International Journal of Aging and Human Development, 84*(3), pp.231–246.

Fleming, N., Ng, N., Osborne, C., Biederman, S., Yasseen III, A. S., Dy, J. and Walker, M., 2013. Adolescentpregnancy outcomes in the province of Ontario: a cohort study. *Journal of Obstetrics and Gynaecology Canada,35*(3), 234–245.

Golden, N.H., Schneider, M. and Wood, C., 2016. Preventing obesity and eating disorders in adolescents. *Pediatrics, 138*(3), p.e20161649.

Gorrell, S. and Le Grange, D., 2019. Update on Treatments for Adolescent Bulimia Nervosa. *Child and Adolescent Psychiatric Clinics, 28*(4), pp.537–547.

Gupta, A., Noronha, J.A. and Garg, M., 2018. Dietary intake of macronutrients and micronutrients among adolescent girls: A cross sectional study. *Clinical Epidemiology and Global Health, 6*(4), pp.192–197.

Habib-Mourad, C., Ghandour, L.A., Moore, H.J., Nabhani-Zeidan, M., Adetayo, K., Hwalla, N. and Summerbell, C., 2014. Promoting healthy eating and physical activity among school children: findings from Health-E-PALS, the first pilot intervention from Lebanon. *BMC public health, 14*(1), p.940.

Hail, L. and Le Grange, D., 2018. Bulimia nervosa in adolescents: prevalence and treatment challenges. *Adolescent health, medicine and therapeutics, 9*, p.11.

Hans, K.B. and Jana, T., 2018. Micronutrients in the life cycle: Requirements and sufficient supply. *NFS journal, 11*, pp.1–11.

Jewell, T., Blessitt, E., Stewart, C., Simic, M. and Eisler, I., 2016. Family therapy for child and adolescent eating disorders: a critical review. *Family Process, 55*(3), pp.577–594.

Kathleen, M and Raymond, J. 2016. Krause's Food & the Nutrition Care Process. 14th Edition. Saunders.

Kumar, M.M., AlBuhairan, F., Galagali, P., Docter, A.D., Weiss, A. and Keough, L., 2018. Addressing Nutritional Disorders in Adolescents. *JOURNAL OF ADOLESCENT HEALTH, 63*(1), pp.120–123.

Leme, A.C.B., Thompson, D., Dunker, K.L.L., Nicklas, T., Philippi, S.T., Lopez, T., Vézina-Im, L.A. and Baranowski, T., 2018. Obesity and eating disorders in integrative prevention programmes for adolescents: protocol for a systematic review and meta-analysis. *BMJ open, 8*(4).

Linardon, J., Wade, T.D., De la Piedad Garcia, X. and Brennan, L., 2017. The efficacy of cognitive-behavioral therapy for eating disorders: A systematic review and meta-analysis. *Journal of consulting and clinical psychology, 85*(11), p.1080.

Lock, J., 2015. An update on evidence-based psychosocial treatments for eating disorders in children and adolescents. *Journal of Clinical Child & Adolescent Psychology, 44*(5), pp.707–721.

López-Sobaler, A.M., Aparicio, A., Rubio, J., Marcos, V., Sanchidrián, R., Santos, S., Pérez-Farinós, N., Dal-Re, M.Á., Villar-Villalba, C., Yusta-Boyo, M.J. and Robledo, T., 2019. Adequacy of usual macronutrient intake and macronutrient distribution in children and adolescents in Spain: A National Dietary Survey on the Child and Adolescent Population, ENALIA 2013–2014. *European journal of nutrition*, *58*(2), pp.705–719.

Maatoug J, Msakni Z, Zammit N, Bhiri S, Harrabi I, Boughammoura L, et al. School-Based Intervention as a Component of a Comprehensive Community Program for Overweight and Obesity Prevention, Sousse, Tunisia, 2009–2014. Prev Chronic Dis. 2015; 12:E160.

Mirmiran, P., Sherafat Kazemzadeh, R., Jalali Farahani, S. and Azizi, F., 2010. Childhood obesity in the Middle East: a review. *EMHJ-Eastern Mediterranean Health Journal, 16 (9), 1009–1017, 2010.*

Mokdad, A.H., Jaber, S., Aziz, M.I.A., AlBuhairan, F., AlGhaithi, A., AlHamad, N.M., Al-Hooti, S.N., Al-Jasari, A., AlMazroa, M.A., AlQasmi, A.M. and Alsowaidi, S., 2014. The state of health in the Arab world, 1990–2010: an analysis of the burden of diseases, injuries, and risk factors. *The Lancet*, *383*(9914), pp.309–320.

Nagl, M., Jacobi, C., Paul, M., Beesdo-Baum, K., Höfler, M., Lieb, R., & Wittchen, H. U., 2016. Prevalence, incidence, and natural course of anorexia and bulimia nervosa among adolescents and young adults. *European child & adolescent psychiatry*, *25*(8), 903–918.

Obermeyer, C.M., 2015. Adolescents in Arab countries: health statistics and social context. *DIFI Family Research and Proceedings*, *2015*(1), p.1.

Ozdemir, A. 2016. Macronutrients in Adolescence. *International Journal of Caring Sciences*, 9, (2), 1162–1166.

Rey-López, J.P., Hashem, R., Hamer, M., McMunn, A., Whincup, P.H., Owen, C.G. and Stamatakis, E., 2019. Prevalence of overweight and obesity and associations with socioeconomic indicators: the study of health and activity among adolescents in Kuwait. *Minerva pediatrica*, *71*(4), pp.326–332.

Sysko, R., Hildebrandt, T. and Walsh, B.T., 2019. Pharmacological treatment of eating disorders.

Szabo, K., Piko, B.F. and Fitzpatrick, K.M., 2019. Adolescents' attitudes towards healthy eating: The role of self-control, motives and self-risk perception. *Appetite*, *143*, p.104416.

Unicef.org. 2020. MENA Nutrition. [online] Available at: <https://www.unicef.org/mena/nutrition> [Accessed 3 November 2020].

# Chapter 3
# Addressing Social Determinants of Oral Health Among Adolescents from MENA

**Reham Khaled Abou El Fadl and Haneen Raafat Fathi Mousa**

## 3.1 Introduction

Broadly, people's lifestyle is a reflection of the sociocultural circumstances which configure their behaviors and personal decisions (Green and Kreuter 1990). In the same manner, adolescents' behaviors are shaped by the various environmental contexts in which they live, including family households, neighborhood, and school which, in the long run, impact their health outcomes and quality of life (Institute of Medicine and National Research Council 2011). Some of those determinants such as adequate housing, economic stability, food security, and strong social relationships are defined as positive health factors as they significantly contribute to maintenance of good health status (Szaflarski 2005). Risk factors, on the other hand, are potentially preventable exposures with specific environmental- or lifestyle-related health hazards such as smoking, which are associated with a broad array of health risks. Those factors may have different impacts at different ages during adolescence. Though adolescence is a critical transitional period in an individual's life course, in general, research on broader determinants of health during this decade of life is quite limited (Viner et al. 2012).

R. K. Abou El Fadl (✉) · H. R. F. Mousa
Faculty of Dentistry, Ain Shams University, Cairo, Egypt
e-mail: rehamabouelfadl@dent.asu.edu.eg; haneenraafat@dent.asu.edu.eg



C. Barakat et al. (eds.), *Adolescent Health in the Middle East and North Africa*,
Global Perspectives on Health Geography,
https://doi.org/10.1007/978-3-030-92107-1_3

In the same manner, oral diseases are commonly an outcome of a myriad of an interplaying set of determinants including psychosocial, economic, cultural, and political factors, which substantially impact individuals' care-seeking behaviors and level of commitment to preventive oral hygiene practices. However, in general, conceptions and observations on social determinants of oral health are not adequately explored.

## 3.2 Methodology

A comprehensive literature review was conducted on the following concepts: (i) potential risk factors associated with oral diseases among adolescence and (ii) health policies and approaches for the prevention of oral diseases in this age group. Two electronic databases, i.e., *Web of Science* and *Embase*, were searched, and only English articles were included. The search was not limited by country; however, literature from MENA countries was prioritized. Editorials and commentaries were excluded.

The search strategy utilized both controlled vocabulary and keywords: [Adolescen*OR teen*OR preteen* OR "high school students" OR Underage] AND ["Oral health" OR "dental health" OR "dental disease" OR "oral disease" OR "dental fear" OR "dental avoidance" OR "dental anxiety" OR "dental checkups" OR "dental appointments" OR "dental treatment" OR "dental care" OR "dental services" OR "oral examinations" OR "oral screenings" OR "dental prevention" OR "atraumatic restorative treatment" OR "dental sealants" OR "dental education" OR "dental awareness" OR "dental neglect" OR "dental needs" OR "dental pain" OR "oral pain" OR "sugary diet" OR "cariogenic diet" OR "sugary drinks" OR "carbonated drinks"].

All the references were imported to EndNote, where duplicates were identified and subsequently removed. Titles and abstracts of all entries were analyzed by two reviewers, and only full texts of relevant references were retrieved and reevaluated.

Additionally, the Global Burden of Disease (GBD) 2017 dataset was used to generate epidemiologic data related to smoking among adolescents and youth in the MENA region, using the online tool Tobacco Visualization (Tobacco Viz). In order to acquire data relevant to adolescence and teenage, only the 10–14 and 15–19 age groups were selected.

## 3.3 Findings

Since decades, the Dahlgren-Whitehead "rainbow model" has been used for conceptualization and visual presentation of the relationship between individuals, the surrounding environment, and their health status (Dahlgren and Whitehead 1991). In Fig. 3.1, we adopted this model to envisage the potential determinants of oral

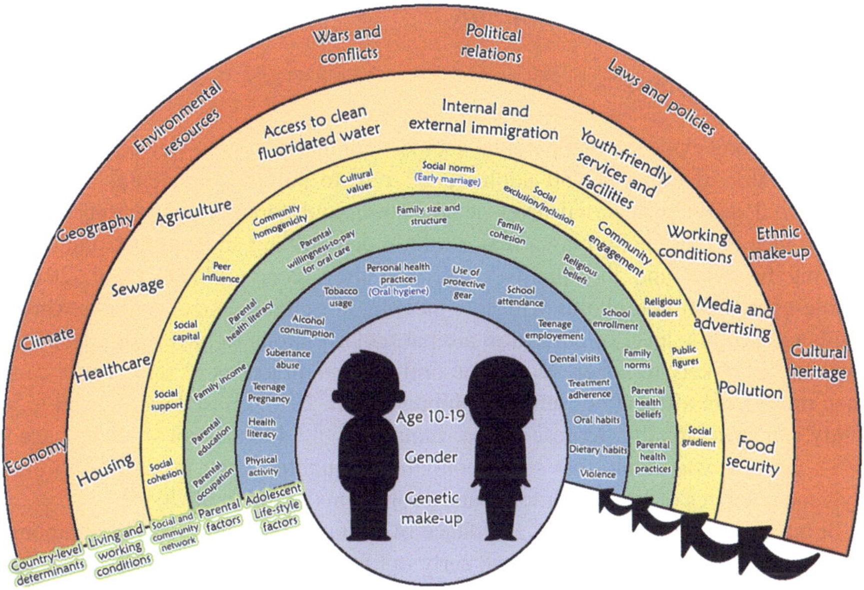


**Fig. 3.1** Social determinants of oral health among adolescents. (Adapted from Dahlgren and Whitehead's model )

health outcomes in youth and adolescence. At the center lie the 10–19-year-olds and the fixed non-modifiable characteristics which influence their health and well-being, such as gender and genetic makeup.

Within the other layers of the model are broader social and economic determinants, which are theoretically modifiable by laws and policies. Given that various family-related factors are strong predictors of adolescents' health outcomes including oral health, the "rainbow model" was modified to include an additional layer representing relevant parental factors, such as family size and structure, parental health beliefs and practices, level of health literacy, and strength of parents-youth relationships. Social media, relationships with school staff, and peer influence were included in the third layer as integral components of social and community networks influencing adolescents' health status and healthcare-seeking behaviors. The fourth layer of the model illustrates different living and working conditions which can directly or indirectly affect access to protective factors such as fluoridated water, healthy and nutritious foods, as well as access to oral health services either in school premises or health facilities. Enforcing high taxation on tobacco products, alcohols, sugar-rich foods, and sweetened beverages; restricting advertisements on those products; enacting laws pertinent to the mandatory use of seatbelts, helmets, or mouth guards; as well as community and school water fluoridation are among the national policies and laws that should be reinforced to promote adolescents' oral health at community rather than individual level.

### 3.3.1 Unpacking Determinants

**Socioeconomic Status**

There is strong evidence that parental socioeconomic position is viewed as the initial indicator of children and adolescents' risk to suffer from dental diseases. Household income, maternal level of education, and parental perceptions of the importance of oral health play a major role in determining adolescents' oral care-seeking behavior and the quality of service they receive (Jacob et al. 2017). Adolescents from households with low income, large family size, or those whose mothers have low level of education are particularly prone to have tooth decay as they are less likely to adhere to routine oral hygiene practices or seek dental care than those from affluent households (Nóbrega et al. 2017). Based on Safiri et al.'s study, on a nationally representative sample of 13,486 school students aged 6–18 years, it was found that socioeconomic inequalities exist in relation to oral hygiene practices of Iranian children and adolescents (Safiri et al. 2016).

It has also been reported that avoidance of dental care is more prevalent among adolescents who were forced to work early in life and those who have a single parent (Gustafsson et al. 2010). Moreover, some health risk behaviors, such as tobacco and substance use, are known to be more common among people of low socioeconomic level (Petersen 2003).

**Tobacco and Substance Use**

Smoking, alcohol drinking, and substance abuse are harmful behaviors that are primarily initiated during adolescence and up until the late 20s (Johnson et al. 2004). It is strongly established that tobacco consumption has various negative effects on oral health. Tobacco is a strong risk factor for precancerous lesions and oropharyngeal cancers and causes staining of teeth and restorations by nicotine, oral candidiasis due to reduced salivary flow, and periodontal problems (Petersen 2003). Among youth, there are various determinants of tobacco consumption, including access to tobacco products, national tobacco control policies, as well as cultural and societal norms. Cigarette advertising and marketing efforts of the tobacco industry highly encourage youth and adolescents to initiate this behavior (Hanewinkel et al. 2010).

It is also worth mentioning that exposure to secondhand smoke whether in households or schools can increase adolescents' risk for defective enamel formation and accordingly caries experience (González-Valero et al. 2018). As per the GBD dataset, in the MENA region, up to 2015, about 4.6 million adolescents aged 10–19 smoke daily, of which 3.8 million are males and 790 thousand are females (Table 3.1). Based on the datasets from the global school-based student health survey (GSHS) in five countries in the MENA region, Fig. 3.2 shows the percentage of 13–17-year-olds who reported exposure to secondhand smoking for one or more days 1 week before the survey (WHO and CDC. GSHS.).

**Table 3.1** MENA countries with highest smoking rates among adolescents

| MENA countries with highest no. of daily smokers[a] | Males | | Females | |
|---|---|---|---|---|
| | 10–14 years | 15–19 years | 10–14 years | 11–15 years |
| #1 | Turkey (130k) | Turkey (800k) | Turkey (60k) | Turkey (350k) |
| #2 | Afghanistan (41k) | Egypt (390k) | Afghanistan (27k) | Afghanistan (95k) |
| #3 | Egypt (35k) | Afghanistan (160k) | Sudan (11k) | Yemen (33k) |

[a]Data was generated from the GBD dataset using Tobacco Visualization (Tobacco Viz) up to 2015 available at https://vizhub.healthdata.org/tobacco/

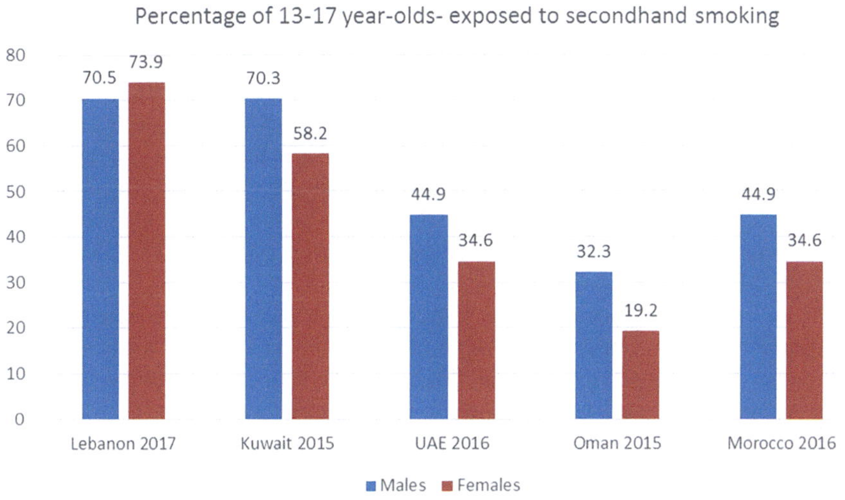


**Fig. 3.2** Exposure to secondhand smoking among adolescents aged 13–17 years. (Source: Global school-based student health survey, (GSHS), WHO in collaboration with CDC https://www.who.int/ncds/surveillance/gshs/en/)

Added to that, lately, the use of smokeless tobacco (ST) has been notably increasing among youth and adolescents. This has been strongly linked to widely spread misconceptions that chewing tobacco or using moist snuff is far less harmful than cigarette smoking (Chaffee et al. 2019). However, ST contains 20-fold more nicotine than a cigarette, which renders users highly prone to oral mucosal lesions, periodontal problems, oral cancers, and premalignant lesions (Percy 2008). Different forms of ST, such as Toombak, Shammah, Qat *(also known as Khat)*, and Nashooq, are commonly used in the Middle East, especially in the Kingdom of Saudi Arabia, Yemen, and Sudan (Alrashidi et al. 2018). Recently, Othman M. et al. conducted the Global Youth Tobacco survey among a representative sample of school-attending adolescents in Khartoum state of Sudan to study the prevalence of smokeless tobacco use. 7.6% of participants reported using smokeless tobacco at least once in their lifetime where male gender, peer influence, exposure to second-hand smoke at home, having single parent and low self-efficacy were strongly associated with this harmful practice. (Othman 2021)

On the other hand, data on substance use was available from the GSHS of only two countries in the MENA region, where 76.7% and 70% of 13–17-year-old drug users in Lebanon and Morocco reported starting before the age of 14. On the other hand, in Lebanon, 13.4% of adolescents in the same age group admitted being fully drunk one or more times in their life.

## Psychosocial Factors

It has been well recognized, from the outset, that psychosocial factors largely affect adolescents' oral health status. During this period, due to various psychological changes, the emerging sense of independence, and peer influence, adolescents are highly vulnerable to harmful behaviors including oral health-related behaviors (Baker et al. 2010). Health-compromising behaviors in adolescence such as smoking, drug abuse, and eating an unhealthy diet are strongly linked to social influence per se, and in various adolescent and adult populations, the combined effects of lack of social support and low numbers of social networks on oral health outcomes, such as dental caries and traumatic oral injuries, have been well-reported (Baxevanos et al. 2017; Fontanini et al. 2015).

On the other hand, social connectedness through listening, appreciating, and expressing love has been proposed as an efficient protective factor for health and well-being. According to (Viner et al. 2012), having safe and supportive families, connectedness with schools, and positive peer influence are among the strongest determinants for the development of adolescents and attainment of best health outcomes in the transition to adulthood. In the same manner, strong social ties and family cohesion can yield significant oral health benefits among adolescents by favoring engagement in health-promoting lifestyle behaviors and regular dental attendance (Camara et al. 2017; Sisson 2007). One study revealed that children from poorly functioning families are more likely to experience dental caries being less likely to engage in oral health-related behaviors or seek dental care (Duijster et al. 2014).

In one study in Saudi Arabia, oral health-related practices, including smoking, tooth brushing, and snacking on sugary foods and drinks, were found to be strongly associated with close friends' rather than parents' or distant peers' practices (El Tantawi et al. 2017). According to the GSHS dataset, a varying percentage of adolescents (ranging from 3.4% to 11.2%) in different MENA countries reported not having any close friends (Fig. 3.3), a finding which according to (Gomes et al. 2020) might adversely affect their oral health-related quality of life due to low self-esteem, lack of sense of belonging and care, as well as erroneous oral health beliefs.

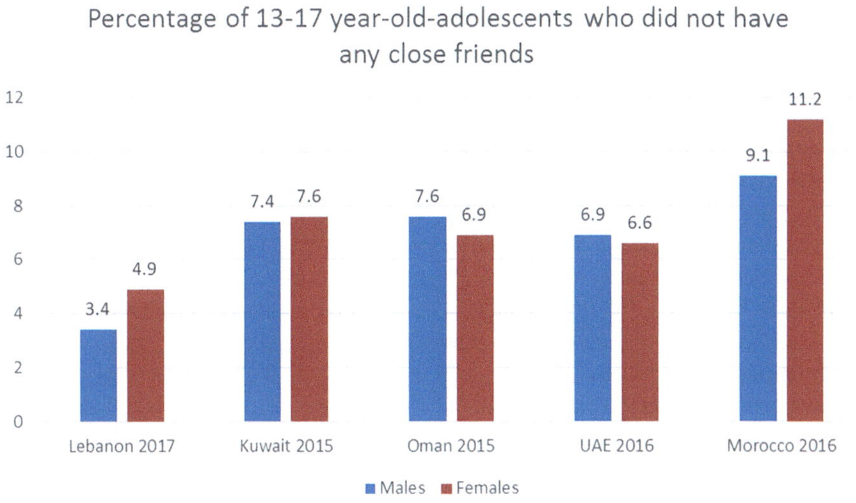


**Fig. 3.3** Absence of close friends among 13–17-year-olds. (Source: Global school-based student health survey, (GSHS), WHO in collaboration with CDC https://www.who.int/ncds/surveillance/gshs/en/)

## 3.4 Discussion

Most research, in the MENA region, is still focused on studying the association between adolescents' lifestyle behaviors and their oral health status and related quality of life. On the other hand, the complex relationships between demographic characteristics, socioeconomic level, and the possible role of social ties and living conditions on the distribution of oral diseases among adolescents have not been well-elucidated.

Adolescence offers a unique opportunity for prevention acknowledging that individuals' risks to suffer from oral disease and other common chronic conditions later in life are heightened if specific risky behaviors start in early adolescence (Gore et al. 2011). In line, the World Health Organization (WHO) has recommended three age groups (12, 15, and 15–19) during this period for conducting oral health surveys. At the age of 12, almost all permanent teeth, except the third molars, would have erupted, and thus a reliable sample could be easily obtained for surveillance of disease trends before children leave primary schools. Measuring caries prevalence at 15 years is also warranted since, by this age, permanent teeth would have been exposed to the oral environment for 3–9 years, while the 15–19 years age group provides a chance for reliable assessment of periodontal health status among adolescents (World Health Organization 2013). Despite that, to date, only few youth-focused oral health programs targeting both salient common risk and protective factors are being implemented in the MENA countries.

### 3.4.1 *Tackling Social Determinants*

**Modifying School Environments**

It is well known that the living environment is a fundamental factor in shaping adolescents' development and schools are the ideal setting for tackling most common health issues including oral diseases. Schools provide excellent outreach opportunities where an unprecedented number of adolescents could benefit from health interventions delivered on its premises. According to the *Lancet Commission on Adolescent Health and Wellbeing* (Patton et al. 2016), health interventions could be delivered across six interdependent platforms including health services, schools, communities, m-health, media and social marketing, as well as structural actions. Though schools cannot replace health facilities for the delivery of health services, clinical care including both preventive and curative oral health services can be delivered in either school-based or mobile dental clinics (Patton et al. 2016). On the other side, there is strong evidence that m-health, which comprises the use of mobile and wireless devices such as mobile phones for tracking and promoting people's health (World Health Organization 2011), can adjunctively improve oral health-related knowledge and practices among people in different age groups (Toniazzo et al. 2019).

For decades, it has been estimated that, annually, over 51 million school hours are lost due to some dental problem (Gift et al. 1993), and thus good oral health status has been viewed as an essential input required for adequate learning. In consideration of the strong link between health and education, in 2002, the FRESH (Focusing Resources on Effective School Health) framework was developed by experts in the United Nations Educational, Scientific and Cultural Organization (UNESCO), United Nations Children's Fund (UNICEF), WHO, World Bank, and Education International to assist governments worldwide and implement school-based health programs in efficient, effective, and results-oriented manners at the national level. The program includes four main components: *health-related school policies*, *provision of safe water and sanitation*, *skills-based health education*, and *school-based health and nutrition services*, which can be used collectively to identify common health issues in school students and suggest potentially effective interventions for addressing those issues. In addition, the framework proposes other supporting activities that can substantially contribute to successful implementation and sustainability of school-based programs, such as effective partnerships between health and education sectors and community and youth participation (Joerger et al. 2002). Table 3.2 illustrates essential strategies and activities that could be adopted for setting up oral health programs in schools using the FRESH framework.

**Table 3.2** Adopting the FRESH framework for promoting adolescent oral health in schools

| FRESH core components | Action | |
|---|---|---|
| Core component # 1: Health-related school policies | Enforce policies banning tobacco/alcohol consumption on school premises<br>Enforce policies banning sugary foods and drinks on school premises<br>Safe playgrounds and sports fields<br>Enforce effective disciplinary measures against abuse and bullying<br>Affordable fluoride toothpaste | |
| Core component #2: Provision of safe water | Regular access to drinking-water fountains throughout the school<br>School water fluoridation | |
| Core component #3: Skills-based health education | Compulsory cooking classes for both genders<br>Daily supervised tooth brushing<br>Training on sanitation-related behaviors such as hand-washing, and food washing<br>Mandated physical education<br>Peer modeling<br>Role playing | |
| Core component #4: School-based health and nutrition services | Nutrition services | Oral health services |
| | Healthy school meals<br>Frequent diet history analysis and height and weight screening<br>Healthy food choices in canteens and vending machines | Oral health education integration in school curricula<br>Regular oral screening<br>Technology application in oral health promotion (e.g. using Mobile apps to address unhealthy diets and tobacco and alcohol use)<br>Topical fluoride administration, e.g., school-based fluoride rinsing/varnish program<br>School-based sealant program for permanent molars and premolars<br>Curative care, e.g., restorative treatment using atraumatic restorative treatment (ART) approach |
| Supporting activities | Counseling on<br>STIs; HIV/AIDS;<br>Pregnancy;<br>Tobacco, alcohol, and drug use;<br>Bullying and physical and sexual abuse<br>Psychosocial support<br>Involvement of parents in health education<br>Training for school staff<br>Youth-friendly services | |

Source: Authors

## Examples of Youth-Focused Oral Health-Related Initiatives in the MENA Countries

Intervention 1: Telegram – Online Social Media Platform Program, Iran (Scheerman et al. 2020)

***Program Overview*** A theory-based program using an online social media platform (Telegram); an open-access platform available for smartphones which allows users to send messages and receive photographs, videos, and audios. The program consisted of two components: oral health education + behavioral coaching component comprising sending text messages about the potential positive outcomes of dental cleaning and disseminating volitional sheets on oral health behaviors. Mothers were instructed on Telegram channel to coach and monitor their child's oral hygiene practices on daily basis.

***Target Population and Setting:*** Adolescents aged 12–17 years recruited from 30 out of 73 high schools in Qazvin City, Iran, and their mothers who had access to Telegram via a smartphone.

**Outcome(s)**
- Significant increase in adolescents' tooth-brushing practices at the 1- and 6-month follow-ups.
- Involving mothers in an intervention conferred additional benefits for adolescent oral health.

Intervention 2: School Oral Health Program (SOHP) – A National Program, Kuwait (Alsumait et al. 2019; Ariga et al. 2014)

***Program Overview:*** SOHP is a nationwide program which started in 1983 as a joint venture between the Ministry of Health in Kuwait and Forsyth Institute, Cambridge, Mass., USA. The multicomponent intervention consisted of the following:

At least two oral health education sessions with supervised tooth brushing during every school year; + oral health education session to parents and expectant mothers + twice-a-year application of fluoride varnish + (if eligible) fissure sealants in newly erupted permanent molars and premolars + restorative treatment of permanent molars when indicated. Delivery of care was performed by almost 207 dentists, 335 nurses, and 25 dental hygienists.

***Target Population and Setting:*** Children and adolescents aged 6–16 years in school-based centers in all six governorates in Kuwait.

**Outcome(s)**

- Overall, children and adolescents enrolled in SOHP had better dental health and lower caries experience.
- There has been a significant reduction in the number of composite fillings performed as part of the SOHP.
- Significant increase in the number of preventive procedures; sealants placement and fluoride applications performed.

## 3.5 Key Recommendations and Future Implications

- Development of adolescent-specific *Oral Health Surveillance plan* including the following:
  - (i) Conducting oral health needs assessment for adolescents, at regular basis, to capture information about their health status, unmet dental needs, and patterns of oral care-seeking behavior using standardized survey tools
  - (ii) Conducting longitudinal studies to understand the unique causal factors of oral diseases among adolescents and explore the synergetic effects of different oral health determinants with particular emphasis on the social environment in which adolescents are embedded
  - (iii) Developing locally driven, culturally appropriate interventions addressing broader health determinants with particular emphasis on factors that are protective across various health outcomes
- Controlling oral diseases in adolescence by adopting "the life course approach" to health through the following:
  - (i) Reorienting oral health services with a focus on preventive interventions early in childhood is a key strategy for halting processes that increase the risk of poor oral health in adolescence and later on.
  - (ii) Incorporating adolescents' oral health into pregnancy, neonatal, and children health issues and interventions.
  - (iii) Pregnant adolescent girls should, also, be entitled to receive all medically necessary dental services as part of basic package of essential services acknowledging that "pregnancy-related" oral care comprises also services for "other conditions which might complicate the pregnancy."
  - (iv) Advocating for using a "whole-school approach" to oral health promotion and adapting the FRESH framework to fit the local context to ensure appropriateness of delivered school health services.
- Integrating youth-specific interventions at school, family, as well as community levels to avoid exclusion of out-of-school adolescents
- Building capacities of dental practitioners to deliver oral care to individuals with special health care needs aided by interprofessional teams including nutrition-

**Table 3.3** Suggested oral health indicators in adolescence and teenage

| | |
|---|---|
| 1 | Number of school hours lost among adolescents due to oral diseases/year |
| 2 | Number of working hours lost to seek oral care for adolescents/year |
| 3 | Proportion of 10–14-year-old school students with dental sealants on at least one permanent molar tooth |
| 4 | Proportion of adolescents aged 10–14 years with untreated dental decay in their permanent teeth |
| 5 | Proportion of adolescents aged 15–19 years with untreated dental decay in their permanent teeth |
| 6 | Proportion of adolescents aged 10–19 who received dental treatment in the past year |
| 7 | Proportion of adolescents with special health care needs who sought dental care in the past year |
| 8 | Proportion of adolescents who have regular access to a toothbrush and fluoridated toothpaste |
| 9 | Proportion of adolescents who never visited a dentist |
| 10 | Number of adolescents with access to school water fluoridation programs |

Source: Authors

ists, social workers, and other non-dental health professionals to optimize the quality of care
- Adopting the "Health in All Policies" approach to advocate for the inclusion of oral health within the mandate of other sectors, such as educational, social, and industrial sectors
- Building strong partnerships with key stakeholders such as civil society, local authorities, community leaders, school districts, academic institutions, and local dental health practitioners to plan for policy priorities pertinent to adolescents' oral health
- Researching the role of non-dental health professionals such as school nurses, midwives, and community health workers in delivering educational and preventive oral health services to adolescents inside and beyond educational settings
- Developing a set of reliable and age-appropriate oral health indicators and including them in regular disease surveillance for rigorously monitoring and evaluating the impact of various social and behavioral interventions on oral health outcomes in this age group (Table 3.3).

## 3.6 Conclusions

National health policies of MENA countries should be reoriented to integrate oral health research into general health research programs and adopt the common risk factor approach for implementing holistic interventions that promote the oral and general health of adolescents as well as other population groups. To achieve that, health interventions should be directed toward disease prevention and should be culturally relevant and match adolescents' unique health needs.

## Reflection Questions

- Among youth and adolescents, what are the key sources of tobacco? And how can we prevent or delay the onset of their tobacco and substance use?
- In what way(s) can peers and families influence oral health of adolescents?
- How can social media impact adolescents' oral health practices?
- Why are schools excellent venue for promoting oral health of school-aged children and adolescents?
- How are governments entitled to ensure equitable access to oral care among adolescents living in disadvantaged communities?

## References

Alrashidi, A.G., Alrashidi, T.G., Alrashedi, S.A., Alreshidi, N.A., Alrashidi, M.F., Alrashidi, A.S., Alkhaldi, E.H., Alrashdi, A.G., Alshammari, M.S., Ahmed, H.G., 2018. Epidemiologic pattern and types of oral smokeless Tobacco usage in Saudi Arabia. J. Contemp. Dent. Pract.

Alsumait, A., ElSalhy, M., Behzadi, S., Raine, K.D., Gokiert, R., Cor, K., Almutawa, S., Amin, M., 2019. Impact evaluation of a school-based oral health program: Kuwait National Program. BMC Oral Health 19, 202.

Ariga, J., Al-Mutawa, S., Nazar, H., 2014. School Oral Health Program in Kuwait. Med. Princ. Pract. 23, 43–46.

Baker, S.R., Mat, A., Robinson, P.G., 2010. What psychosocial factors influence adolescents' oral health? J. Dent. Res.

Baxevanos, K., Topitsoglou, V., Menexes, G., Kalfas, S., 2017. Psychosocial factors and traumatic dental injuries among adolescents. Community Dent. Oral Epidemiol.

Camara, M., Bacigalupe, G., Padilla, P., 2017. The role of social support in adolescents: are you helping me or stressing me out? Int. J. Adolesc. Youth.

Chaffee, B.W., Couch, E.T., Urata, J., Gansky, S.A., Essex, G., Cheng, J., 2019. Predictors of Smokeless Tobacco Susceptibility, Initiation, and Progression Over Time Among Adolescents in a Rural Cohort. Subst. Use Misuse.

Dahlgren, G., Whitehead, M., 1991. Policies and strategies to promote social equity in health. Available: https://core.ac.uk/download/pdf/6472456.pdf.

Duijster, D., Verrips, G.H.W., Van Loveren, C., 2014. The role of family functioning in childhood dental caries. Community Dent. Oral Epidemiol.

El Tantawi, M., Bakhurji, E., Al-Ansari, A., Al-Khalifa, K.S., AlSubaie, A., 2017. Influences of parents, close friends and classmates on four co-existing oral health practices in Saudi male teenagers. Acta Odontol. Scand.

Fontanini, H., Marshman, Z., Vettore, M., 2015. Social support and social network as intermediary social determinants of dental caries in adolescents. Community Dent. Oral Epidemiol.

Gift, H.C., Reisine, S.T., Larach, D.C., 1993. Erratum: The social impact of dental problems and visits (Am J Public Health (1992) 82 (1663-1668)). Am. J. Public Health.

Gomes, A.C., Rebelo, M.A.B., de Queiroz, A.C., de Queiroz Herkrath, A.P.C., Herkrath, F.J., Rebelo Vieira, J.M., Pereira, J.V., Vettore, M.V., 2020. Socioeconomic status, social support, oral health beliefs, psychosocial factors, health behaviours and health-related quality of life in adolescents. Qual. Life Res.

González-Valero, L., Montiel-Company, J.M., Bellot-Arcís, C., Almerich-Torres, T., Iranzo-Cortés, J.E., Almerich-Silla, J.M., 2018. Association between passive tobacco exposure and caries in children and adolescents. A systematic review and meta-analysis. PLoS One.

Gore, F.M., Bloem, P.J.N., Patton, G.C., Ferguson, J., Joseph, V., Coffey, C., Sawyer, S.M., Mathers, C.D., 2011. Global burden of disease in young people aged 10-24 years: A systematic analysis. Lancet.

Green, L.W., Kreuter, M.W., 1990. Health promotion as a public health strategy for the 1990s. Annu. Rev. Public Health.

Gustafsson, A., Broberg, A.G., Bodin, L., Berggren, U., Arnrup, K., 2010. Possible predictors of discontinuation of specialized dental treatment among children and adolescents with dental behaviour management problems. Eur. J. Oral Sci.

Hanewinkel, R., Isensee, B., Sargent, J.D., Morgenstern, M., 2010. Cigarette Advertising and Adolescent Smoking. Am. J. Prev. Med.

Institute of Medicine, National Research Council., 2011. The Science of Adolescent Risk-Taking Workshop Report, Committee on the Science of Adolescent.

Jacob, C.M., Baird, J., Barker, M., Cooper, C., Hanson, M., 2017. The Importance of a Life Course Approach to Health: Chronic Disease Risk from Preconception through Adolescence and Adulthood, Who Report.

Joerger, C., Hoffman, A.M., Aldinger, C., Bundy, D., Fouilhoux, M., Gillespie, A., Jones, J., Vince-Whitman, C., UNESCO, 2002. FRESH: A comprehensive school health approach to achieve EFA 32.

Johnson, C.C., Myers, L., Webber, L.S. and Boris, N.W., 2004. Profiles of the adolescent smoker: models of tobacco use among 9th grade high school students: Acadiana Coalition of Teens against Tobacco (ACTT). *Preventive medicine, 39*(3), pp. 551–558.

Nóbrega, J., Dantas, E., Ferreira-Filho, J.C., Limão, N., Rodrigues-de-Melo, A.C., Protásio, A.P., Valença, A.M., Santiago, B., 2017. Contextual Social Inequities and Occurrence of Dental Caries in Adolescents: A Systematic Review. Oral Health Prev. Dent. 15, 329–336.

Othman, M., Farid, N.D.N., Aghamohammadi, N. and Danaee, M., 2021. Determinants of smokeless tobacco use and prevalence among Sudanese adolescents. *Archives of Public Health, 79*(1), pp.1–10.

Patton, G.C., Sawyer, S.M., Santelli, J.S., Ross, D.A., Afifi, R., Allen, N.B., Arora, M., Azzopardi, P., Baldwin, W., Bonell, C., Kakuma, R., Kennedy, E., Mahon, J., McGovern, T., Mokdad, A.H., Patel, V., Petroni, S., Reavley, N., Taiwo, K., Waldfogel, J., Wickremarathne, D., Barroso, C., Bhutta, Z., Fatusi, A.O., Mattoo, A., Diers, J., Fang, J., Ferguson, J., Ssewamala, F., Viner, R.M., 2016. Our future: a Lancet commission on adolescent health and wellbeing. Lancet.

Percy, M.S., 2008. Oral health of adolescents--it's more than dental caries. MCN. Am. J. Matern. Child Nurs.

Petersen, P.E., 2003. Tobacco and oral health--the role of the world health organization. Oral Health Prev. Dent.

Safiri, S., Kelishadi, R., Heshmat, R., Rahimi, A., Djalalinia, S., Ghasemian, A., Sheidaei, A., Motlagh, M.E., Ardalan, G., Mansourian, M., Asayesh, H., Sepidarkish, M., Qorbani, M., 2016. Socioeconomic inequality in oral health behavior in Iranian children and adolescents by the Oaxaca-Blinder decomposition method: The CASPIAN- IV study. Int. J. Equity Health 15, 1–8.

Scheerman, J.F.M., Hamilton, K., Sharif, M.O., Lindmark, U., Pakpour, A.H., 2020. A theory-based intervention delivered by an online social media platform to promote oral health among Iranian adolescents: a cluster randomized controlled trial. Psychol. Health 35, 449–466.

Sisson, K.L., 2007. Theoretical explanations for social inequalities in oral health. Community Dent. Oral Epidemiol.

Szaflarski M., 2005. The Impact of Inequality: How to Make Sick Societies Healthier. *Preventing Chronic Disease, 3*(1), A26.

Toniazzo, M.P., Nodari, D., Muniz, F.W.M.G., Weidlich, P., 2019. Effect of mHealth in improving oral hygiene: A systematic review with meta-analysis. J. Clin. Periodontol.

Viner, R.M., Ozer, E.M., Denny, S., Marmot, M., Resnick, M., Fatusi, A., Currie, C., 2012. Adolescence and the social determinants of health. Lancet.

World Health Organization, 2011. mHealth: New horizons for health through mobile technologies. Observatory.

World Health Organization, 2013. Oral Health Surveys - Basic Methods 5th edition [WWW Document]. WHO.

World Health Organization (WHO) and Center for Disease Control and prevention (CDC). Global school-based student health survey (GSHS). Glob. Sch. student Heal. Surv. URL https://www.who.int/ncds/surveillance/gshs/en/ (accessed 10.25.20).

# Chapter 4
# Physical Activity and Sport Participation Among Adolescents from MENA

**Caroline Barakat, Susan Yousufzai, Samah Mohammed, and Rania Dghaim**

## 4.1 Introduction

Physical activity (PA) is defined as any bodily movement produced by muscles that requires energy expenditure above the resting metabolic rate (RMR). PA includes non-structured or structured activities, organized sports, or active play. Examples of non-structured activities are leisure walking, cycling, and dancing (WHO 2020). Organized sport is defined as "a subset of PA that is structured, goal-oriented, competitive and contest-based" (Paulo et al. 2018). Active play may involve "symbolic activity or games with or without clearly defined rules; the activity may be unstructured or unorganized, social or solitary, but the distinguishing features are a playful context, combined with activity that is significantly above resting metabolic rate" (Paulo et al. 2018). PA is associated with various health benefits, such as lowering the risk of developing chronic conditions, improving mental health, and increasing overall well-being (Craggs et al. 2011; Eime et al. 2013). Despite the various benefits, many individuals are not physically active; in fact, low levels of PA are main risk factors for deaths around the world and key risk factors for non-communicable diseases, such as cancer, diabetes, and obesity (WHO 2020).

Among young people, regular PA relates to improved self-esteem, academic achievement, and cognitive performance (Chaabane et al. 2020). In addition to these relatively short-term benefits, regular youth participation in PA and sport relates to

C. Barakat (✉) · S. Yousufzai · S. Mohammed
Faculty of Health Sciences, Ontario Tech University, Oshawa, ON, Canada
e-mail: Caroline.barakat@ontariotechu.ca; susan.yousufzai@ontariotechu.ca

R. Dghaim
College of Natural and Health Sciences, Zayed University, Dubai, United Arab Emirates



C. Barakat et al. (eds.), *Adolescent Health in the Middle East and North Africa*, Global Perspectives on Health Geography,
https://doi.org/10.1007/978-3-030-92107-1_4

long-term benefits across the life course. In particular, adolescence is a key period of transition between childhood and adulthood, when the formation and establishment of many lifestyle patterns and behaviors begins and evolves across the life course. The World Health Organization defines an adolescent as an individual between the ages of 10 and 19. Regular PA and sport participation during adolescence can have implications on one's health in later life by increasing confidence, self-efficacy, the sense of belonging, and improved health (Allafi et al. 2014; Sarsour et al. 2019; Tesler et al. 2019), all of which can promote continuous PA and/or sport participation across the life course.

Metabolic equivalent (MET) values are measures that indicate whether the PA is light intensity, moderate intensity, or vigorous intensity. MET allows for precise measurement of both total energy expenditure (EE) and energy intake (EI) based on time spent in various intensities of activity (Ridley et al. 2008). The MET intensity values have been developed as part of an instrument known as "The Compendium of Physical Activity" for adults and its equivalent for youth (Al-Hazzaa et al. 2011; Hamrani et al. 2015). For example, vigorous-intensity PA include activities such as, stair-climbing, jogging, running, cycling, self-defense, weight training, and vigorous sports, such as soccer, basketball, handball, and singles tennis (8 METs), while low-intensity activities may include household chores (3 METs) (Al-Hazzaa et al. 2011). The global consensus for the recommended amount of moderate-to-vigorous-intensity PA(MVPA) for health in children and adolescents aged 5–18 years is a minimum of 60 min daily (Paulo et al. 2018; Tremblay et al. 2011). For adults between the ages of 18 and 64 years, the recommended guidelines are a minimum of 150 min of moderate aerobic PA or at least 75 min of vigorous intensity of aerobic throughout the week (Chaabane et al. 2020; Tremblay et al. 2011). Undertaking more than the minimum daily amount of PA incurs additional health benefits, and incorporating aerobic vigorous-intensity activities sustains and increases muscle strength (WHO 2020).

The importance of PA promotion is more salient in geographical regions that exhibit environmental factors that pose challenges to adopting recommended PA guidelines, particularly if these factors are modifiable. For example, some places may promote behavioral and societal practices for younger generations (particularly girls), which are mostly based on sedentary activities rather than on moderate-to-vigorous PA (Tesler et al. 2019). Other places may hinder PA due to physical environmental features, such as loss of green space and open fields due to urbanization or low access to sport facilities. For many countries in the Middle East and North Africa (MENA), research findings consistently suggest that low levels of adolescent PA are public health concerns (Chaabane et al. 2020; Yammine 2017). Political and economic landscapes across MENA countries appear to hinder adolescent participation in sport and physical activity, due to a number of geographically based factors, such as low access to resources and opportunities, particularly in less affluent countries (Tesler et al. 2019). For example, in the Gaza Strip, social and environmental determinants interact to hinder PA levels in various communities; these determinants include insufficient public areas for practicing physical activity, the negative effects of the unstable political situation on the behavior of the community, and low awareness on the importance and benefits of PA (Sarsour et al. 2019).

In addition to the influence of place-specific socioeconomic and physical features on PA and sport participation levels, individual factors play a big role in influencing adolescent participation in sport and physical activity. Indeed, a multitude of individual, communal, social, physical, and cultural factors interact to influence adolescent sport and PA levels. These include specific gender roles, cultural beliefs, environmental features, and individual experiences (Sharara et al. 2018; Tesler et al. 2019). In particular, research findings have long documented the influence of gender factors on adolescent sport and PA participation. Girls are more likely to have negative experiences in relation to sport and PA than boys, which hinders their interest in sport and PA and leads to less involvement in subsequent leisure time PA (Tesler et al. 2019). Specifically, girls may experience negative effects due to their perceived competence in relation to sport and physical activity, which often stems from being negatively assessed by their coaches, teachers, or peers (Klicnik et al. 2020; Mohammed et al. 2021).

### *4.1.1 Physical Activity and Sport Participation in the MENA*

A recent study that explored PA in the MENA region highlighted that only approximately 25% of youth were sufficiently active (Chaabane et al. 2020). Majority of adolescents fail to meet the recommended PA guidelines of 60 min of PA per day (Musaiger et al. 2013; Sharara et al. 2018; Yammine 2017). For example, UAE-based research has found that the prevalence of adolescents who meet the recommended medium-vigorous PA has fallen from 20% to 16% over a 13-year time period (Paulo et al. 2018; Zaabi et al. 2016).

Low levels of PA are positively associated with heart disease morbidities, childhood obesity, and other chronic diseases in the MENA (Benajiba et al. 2020). Factors that contribute to low levels of adolescent PA include heavy reliance on modern transportation rather than active transport options and spending long periods of screen time including watching television or surfing the Internet (Benajiba et al. 2020; Musaiger et al. 2013; Sharara et al. 2018). Indeed, research findings highlight high levels of sedentary behavior (less than 1.5 metabolic equivalents), where adolescents' behaviors exceed the recommended screen time per day guidelines (Henry et al. 2004; Paulo et al. 2018). Given the short- and long-term benefits of PA during adolescence, it is crucial to investigate factors that influence PA levels and sport participation among adolescents from the MENA. *Barriers* or *constraints* to PA or sport participation are factors that hinder PA levels. This chapter will refer to these factors as *constraints* to emphasize that factors can be mitigated or negotiated rather than act as conclusive *barriers* to PA and sport participation. Comprehensive knowledge on constraints to PA and sport participation can facilitate culturally appropriate interventions and has the capacity to inform programming and policy that promote healthier behaviors among young people (Obermeyer et al. 2015; Chaabane et al. 2020).

This chapter synthesizes evidence on the prevalence and constraints to PA and sport participation among adolescents from the MENA. The chapter uses a case study approach to examine results of a brief study on the prevalence of sport participation among adolescents from the UAE. The chapter objectives are to present evidence related to the prevalence of PA among adolescents from the MENA, to identify constraints to PA for these adolescents, and to examine the prevalence of PA and sport participation among adolescents from the UAE.

## 4.2 Methodology

### *4.2.1 Literature Review*

A literature search used four main databases: CINAHL, PubMed, Google Scholar, and Ovid. The following terms were used in each database: (1) "physical activity" or "exercise" or "fitness" or "sport," (2) "adolescent*" or "teen" or "youth" or "young adults," and (3) "Middle East," "MENA," "Arab," or "UAE." The searches were limited to articles published in English between 2000 and 2020. Two researchers (SY and SM) screened the retrieved articles using a multistep process, first using the article titles followed by the abstracts. Articles were included in this review if they addressed PA among adolescents in the MENA. Studies that did not focus on adolescents or occurred outside the MENA were excluded. Researchers examined articles that passed the title and abstract screening and extracted data related to prevalence of PA and related constraints, using a self-developed data extraction sheet.

### *4.2.2 Case Study UAE*

#### Setting and Study Design

Case study data were obtained from the National Study of Population Health in the UAE (*NSPHUAE)* research program, conducted between 2007 and 2009 (Barakat-Haddad 2013). This program was carried out in collaboration with the UAE Ministry of Education (MOE) and was based on cross-sectional data from 147 private and public schools in the 7 emirates, distributed across 9 educational zones. A survey was developed and administered to 6363 adolescents aged 13–20, who attended the selected schools.

**Analysis**

Data analysis related to select survey items that focused on PA and sport participation. Analyses focused on the prevalence of ever participating in PA or sport in the last 12 months, as well as responses to the following items: "I do physical exercise" (to maintain my health) (yes or no) and "which of the following physical activities have you practiced in the past year during your leisure time?" Two follow-up questions asked whether these activities occurred indoors or outdoors and the duration of these activities. Response options included walking for exercise, swimming, bicycling, popular or social dance, home exercises, skating or rollerblading, jogging or running, golfing, exercise class or aerobics, bowling, tennis, weight training, fishing, volleyball, basketball, soccer, and an open-ended option for other sports or forms of physical activity. IBM SPSS 27 generated univariate frequency analysis. Types of PA were categorized as either organized (structured) sport or unstructured or active play; the latter included walking, swimming, bicycling, home exercise, skating or rollerblading, jogging or running, exercise class, bowling, weight training, and fishing. Organized or structured sport included tennis, volleyball, basketball, soccer, dancing, and golfing. In order to calculate the proportion of participants who report being outdoors for several hours a day, the number of hours per weeks for participants who reported participating in activities outdoors was aggregated. A total that exceeded 7 h a week approximated whether participants reported being outdoors for several hours a day. Chi-square analysis determined potential gender differences in PA and sport participation.

## 4.3 Results

### *4.3.1 Literature Review*

The search strategy led to the retrieval of 17 research articles that focused on PA levels among adolescents from the MENA. One article reported on a systematic review (Sharara et al. 2018), one was a meta-analysis on five UAE-based studies (Yammine 2017), and one was cross-cultural (Musaiger et al. 2013). Two original studies took place in Palestine (Al Sabbah et al. 2007; Sarsour et al. 2019), and three studies were based in Israel (Kaluski et al. 2009; Tesler et al. 2019; Yaffe 2018). Four studies were UAE-based (Bani-Issa et al. 2020; Henry et al. 2004; Paulo et al. 2018; Wasfi et al. 2008); two studies were based in the Kingdom of Saudi Arabia (Al-Hazza et al. 2011; Alsubaie and Omer 2015), and one study was in each of Morocco (Hamrani et al. 2015), Lebanon (Fazah et al. 2010), and Kuwait (Allafi et al. 2014). For most studies, recruitment consisted of school-aged adolescents with sample sizes varying from 58 to over 16,000 participants. Most studies were cross-sectional and employed quantitative analyses. Extracted themes focused on gender and demographic differences in relation to PA levels and diverse constraints to PA and sport participation.

## Gender and Demographic Differences in Relation to Physical Activity

A systematic review based on 172 studies reported on physical inactivity levels for adolescents from the MENA (Sharara et al. 2018). Physical inactivity among children or adolescents was based on nationally representative samples from the global school-based student health surveys (GSHS) ($n$ = 46,426) and was defined as participation in PA for less than 60 min per day on 5 or more days during the past 7 days. Results from the GSHS suggested that physical inactivity is relatively high for MENA adolescents, ranging from 65% in Lebanon to 91% in Egypt. Synthesis from original research studies ($n$ = 20) confirmed high levels of physical inactivity and low levels of PA and sport participation. Overall, the prevalence of inactivity was higher among women or girls for 161 of the 172 studies that were included in the review.

Similarly, a systematic review and meta-analysis of five prospective studies ($n$ = 12,782) that focused on mild, moderate, and vigorous PA among adolescents from the UAE found that girls were more likely to engage in mild physical activity, while moderate and vigorous PA were significantly higher among boys (Yammine 2017). In addition, approximately 19% of Emirati students compared to 22% of expatriate students never participated in physical activity. Results from the United Arab Emirates' 2018 Report Card on PA reported that while overall 16% of UAE children met the recommended guidelines for moderate-vigorous physical activity, expatriate children and boys had higher levels of PA compared to Emirati children and girls (Paulo et al. 2018). Expatriate children were more physically active (total 17%; M 22%; F 12%) compared to Emirati children (total 14%; M 18%; F 10%) across the examined age groups and gender. Furthermore, the study reported that PA declined from ages 13–15 years (total 17%; M 22%; F 12%) to ages 16–17 years (total 14%; M 19%; F 10%). An earlier UAE-based study reported that the prevalence of PA was higher (33.9%) among Emirati private secondary school adolescents compared to expatriates (34% versus 19%) in relation to vigorous exercise for more than three times per week for 20 min. Significant gender differences also were seen for boys (26%) compared to girls (15%) (Wasfi et al. 2008). In addition, regular sport participation was significantly higher among Emirati students (51%) compared to their expatriate adolescents (40%). A cross-sectional study ($n$ = 58) that examined physical activity levels (PAL) and the activity-related energy expenditure among adolescent girls from Abu Dhabi, UAE (using a 3-day activity diary), reported relatively low PAL compared to other countries (e.g., PAL of 1.26 versus 1.73–1.80 for Sweden).

Gender differences in PAL were also reported among adolescents who reside in the West Bank, Gaza Strip, and Israel. A large cross-sectional study ($n$ = 8885) that examined PAL for adolescent students from the West Bank (53%) and the Gaza Strip (47%) found that boys were significantly more physically active for more than 5 days per week than girls ($p < 0.01$), while girls reported doing more homework ($p < 0.001$) (Al Sabbah et al. 2007). Physical activity levels declined with increasing age for both boys and girls ($p < 0.001$). In addition, PAL were higher among adolescents who reside in the West Bank compared to those who reside in the Gaza Strip

(25% versus 13%, $p < 0.001$). In a more recent study, Sarsour et al. (2019) examined levels of PA and sedentary behavior among 205 adolescent boys and 173 adolescent girls who reside in Palestine. The study found that 30% of adolescents were classified as active, with significant gender differences (6.4% of girls versus 50% of boys). In addition, 86% of girls had a significantly higher frequency of sedentary behavior than boys (34.1%, $p < 0.001$). The study also reported significant gender differences in the mean number of minutes per day of PA, with 58 min for boys versus 68 min for girls. A WHO-based study ($n$ = 16,145) reported significant sociodemographic differences in leisure physical activity levels among Arab adolescents (87%) and those from Jewish decent (73%, $p < 0.01$) (Tesler et al. 2019). Among boys, lack of PAL was 17% for Arab adolescents compared to 13% for adolescents of Jewish decent, while 25% of adolescent Arab girls did not engage in PA compared to 23% of their counterparts.

Gender differences in PAL were also shown among adolescents from the Kingdom of Saudi Arabia. Al-Hazzaa et al. (2011) examined PA levels using MET values for 2908 secondary school students from three major cities in Saudi Arabia (Riyadh, Jeddah, and Al-Khobar). Vigorous-intensity sports were assigned an average MET value of 8, while moderate and household activities were assigned an average of 4 and 3 or lower, respectively. Results suggested that 56% of adolescent boys and 22% of girls met the current recommendations of 1 h daily of moderate-intensity PA. Compared with boys, girls on average were significantly ($p < 0.05$) more sedentary (6.6 versus 5.3 h/week for combined TV time and computer use) and less active (1211.1 versus 3051.4 METs-min/week), especially with vigorous-intensity physical activities (554.4 versus 2149.9 METs-min/week). In terms of minutes per week, the total physical activity time for adolescent boys was 503.3 ± 475.4 and 265.9 ± 313.0 min for girls, which equates to about 72 and 38 min of daily PA for adolescent boys compared to girls.

A study based in Lebanon ($n$ = 1000) examined PA levels using the MET system by means of the compendium of PA (Fazah et al. 2010). Results were expressed in MET/week for PA in school, health clubs, and leisure time and at home. The study found significant gender differences in PA levels with adolescent boys reporting higher total scores than girls for the normal-weight, overweight, and obese groups ($p = 0.0001$, $p = 0.004$, and $p = 0.0024$, respectively). For example, scores for PA at health clubs were higher for adolescent boys compared to girls in the normal-weight group (16.33 ± 31.9 versus 4.39 ± 12.3); for school PA was 7.17 ± 8.7 for boys compared to 5.45 ± 3.6 for girls; the same was seen for PA at home with boys reporting 24.00 ± 29.7 compared to 21.12 ± 29.0 for girls.

Similarly, a Kuwait-based study ($n$ = 463 boys and 443 girls) reported that 45% of adolescent boys and 76% of adolescent girls get the recommended daily PA level of more than 2520 MET-min per week (Allafi et al. 2014). Boys were significantly more physically active as determined by the total MET-min score per week ($p = 0.001$). Moreover, 96.3% of adolescent boys and 96.7% of girls reported sedentary behaviors, with girls significantly spending more time per day watching television ($p = 0.02$) and using a computer ($p < 0.001$).

A similar study examined PA behaviors for 669 randomly recruited adolescents from secondary schools in Kenitra, Morocco (Hamrani et al. 2015). Participants were categorized into physically active or inactive based on total PA cutoff scores of 1680 MET-min/week (60 min/d × 7 d/week×4 MET), corresponding to 1 h of daily moderate-intensity PA (Hamrani et al. 2015). Overall, one in five adolescents were inactive, with almost 45% reporting television viewing for more than 2 h per day, and 38% engaged in computer use for a similar period. The study found that adolescent boys were more active than girls across a typical week and engaged in more vigorous-intensity PA (such as jogging, running, soccer, basketball) than adolescent girls, who instead spent more time undertaking moderate-intensity PA (e.g., normal-pace walking, brisk walking, recreational swimming, household activities, and sports such as volleyball, badminton and table tennis). The study also reported that most adolescent girls (67%) engaged in PA at school with classmates.

## Constraints to Physical Activity in the MENA

A recent systematic review examined evidence from 23 studies in the MENA that focused on constraints to PA among adolescents (Sharara et al. 2018). Reported constraints included gender (F); increasing age; low socioeconomic status; lack of support and encouragement from parents, peers, and teachers (reported in 16 studies); hot climate (9 studies); attitudes in relation to PA such as insufficient motivation or interest (8 studies); lack of knowledge on the benefits of physical activity; and insufficient time (17 studies). In addition, 23 studies highlighted the built environment as a constraint to physical activity, and 8 studies reported constraints related to gender norms, such as the norms of dressing conservatively not being suitable for physical activity, the need for girls to be chaperoned in public spaces, and the scarcity of gender-segregated fitness facilities.

A cross-cultural study ($n$ = 4698) across seven countries from the MENA reported similar constraints related to gender (F), lack of motivation, low support from teachers, time, culture, religion, and lack of opportunities for girls (Musaiger et al. 2013). Overall, across all countries included in the study (Algeria, Jordan, Kuwait, Libya, Palestine, Syria, and the United Arab Emirates), girls faced more barriers to PA than boys (Musaiger et al. 2013). Importantly, significant gender differences existed for lack of motivation (ranging from 13% to 19% for girls compared to 6–20% for boys across the seven countries), not having teacher support in relation to participation in PA (ranging from 21% to 36% for girls compared to 16–39% for boys), and time (range of 39–68% for girls compared to 29–45% for boys).

Studies based in the UAE also highlighted diverse and similar constraints to PA and sport participation, including age, gender (F), smoking status, father's education level, socioeconomic status, school workload, unavailability of places for sport participation, high costs of sport clubs, climate, transportation, school and parental support, culture, and low self-efficacy (Henry et al. 2004; Bani-Issa et al. 2020; Wasfi et al. 2008; Yammine 2017). According to results from the United Arab

Emirates' 2018 report on PA for children and youth, PA levels declined from early to late adolescence (aged 13–17 years) among both Emirati and expatriate boys and girls (Paulo et al. 2018). Particularly for adolescent girls, poor levels of PA were related to factors including cultural and weather restrictions and social change of the community (Henry et al. 2004). Wasfi et al. (2008) found significant positive associations between low levels of adolescent sport practice and each of obesity and tobacco smoking, the latter only significant for adolescent boys. Furthermore, 22.3% of average-weight students undertook vigorous exercise for more than three times per week for about 20 min per session and also more than 30 min of moderate PA most days of the week, compared to 18% of students who classified as obese. Among adolescent smokers, 62% of boys had no or a poor level of sport practice compared to 70% of girls. Perceived exercise self-efficacy, defined by the belief in one's ability to maintain an exercise routine, was significantly higher among adolescent boys than girls (t (608) = 5.31, $p < 0.001$) (Bani-Issa et al. 2020). However, UAE adolescents had lower perceived self-efficacy compared to adolescents from Western countries. Significant predictors of exercise self-efficacy for UAE adolescents were age, city location, eating breakfast regularly at home, school support in exercising regularly, and how often schools offered physical exercise classes, with frequency of weekly exercise classes emerging as the strongest predictor of perceived exercise self-efficacy ($\beta = 0.183$, $p = 0.001$) (Bani-Issa et al. 2020).

Consistently, increasing age emerges as a constraint to PA for studies conducted in Palestine and Israel, particularly for girls (Al Sabbah et al. 2007; Kaluski et al. 2009; Sharara et al. 2018; Tesler et al. 2019). Earlier studies also reported that PA levels were related to gender norms, location, overcrowding, lack of sport centers, maternal education, smoking, parental education, socioeconomic status, school workload, high costs, weather, transportation, and urban changes (Al Sabbah et al. 2007; Kaluski et el. 2009). Al Sabbah et al. (2007) found that boys were significantly more physically active than girls but that PA levels significantly decrease with increasing age for both boys and girls. Participation in PA for Palestinian adolescents was positively associated with maternal level of education (OR: 1.26, CI 1.09–1.46, $P < 0.01$). The same study concluded that decreasing levels of PA with increasing age for boys may be linked to social responsibility to secure a future and income, particularly given the overcrowded living conditions and low standards of living in some places (Al Sabbah et al. 2007). The study found that girls reported doing more homework than boys ($p < 0.001$) in Palestine and watching more TV than boys in the West Bank ($p < 0.001$). Kaluski et al. (2009) found that levels of sport practice were associated with both maternal and paternal education (OR = 0.69, 0.58–0.84, 95% CI). Age also emerged as a constraint to sport practice, with 53% more girls from middle school classified as optimally physically active compared to girls from high school ($p = 0.001$). The study concluded that PA levels decline with increasing responsibilities in relation to higher education. For boys, PA levels were negatively related with sleep time for 6 h per night (OR = 0.56, 0.43–0.73 95% CI), smoking hookah, and socioeconomic status. Specifically, boys from lower social status schools were 1.29 (95% CI 1.06, 1.56) times more likely to be optimally active than boys from higher SES schools (OR = 1.30; 95% CI 1.02, 1.65). The type of involvement in sport was related to gender, with more boys reporting involvement in ball games, such as tennis, football, basketball, and volleyball, while girls reported participating more commonly in walking or aerobic activities.

More recent studies reaffirmed PA constraints related to gender, culture, age, parenting style, exercise self-efficacy, peer influence, costs, and parental and school support (Yaffe 2018; Sarsour et al. 2019; Tesler et al. 2019). Findings suggest that boys are given less constrictions and more opportunities to freely practice PA outside their homes in the Gaza Strip (Sarsour et al. 2019). Parents who participated in focus group sessions highlighted that boys are more interested in PA than girls. Quantitative analyses confirmed these results as 86% of girls were involved in sedentary behavior compared to 34% among boys. Gender differences in relation to interest in PA was reported in other regional studies, often due to girls' negative experiences, low motivation, and low perceived self-efficacy (Sharara et al. 2018; Tesler et al. 2019). In addition to age and gender, Tesler et al. (2019) reported that parental engagement in PA was positively related to PA among girls (OR = 0.79; $p < 0.05$). Family affluence was also closely related to parental education and further influences PA levels. Tesler et al. (2019) reported positive associations between PA and family wealth for both girls and boys (OR = 0.92, $p < 0.1$; OR = 0.93, $p < 0.01$, respectively). Significant effects for parenting styles on PA levels were also observed, such that Arab adolescent boys who perceived their parents as authoritative reported doing more PA than their counterparts who perceived their parents as either permissive or authoritarian (Yaffe 2018). Tesler et al. (2019) found that having friends engaged in PA was associated with greater lower likelihood of physical inactivity (OR = 0.64, $p < 0.05$). Additionally, the existence of an in-school physical exercise break was associated with a lower chance of lack of PA (OR = 0.76, $p < 0.01$), and a positive attitude toward physical exercise was inversely related to lack of PA (OR = 0.77, $p < 0.01$).

Consistently, across different countries from the MENA, studies reported diverse constraints to PA. For example, studies based in the Kingdom of Saudi Arabia found evidence of constraints related to gender, age, perceived body weight, absence of sport facilities, lack of peer support, and lack of suitable sport club in the community (Alsubaie and Omer 2015; Al-Hazzaa et al. 2011). Studies based in Morocco, Lebanon, and Kuwait reaffirmed PA constraints related to gender, body weight, and types of PA.

### 4.3.2 Case Study on Physical Activity and Sport Participation in the UAE

Results of the National Study on Population Health in the United Arab Emirates ($n = 6363$) (NSPHUAE) found that UAE adolescents participate in various forms of PA and organized sport (Figs. 4.1 and 4.2). Table 4.1 presents results of data analysis related to PA and sport participation of UAE adolescents. Overall, 73% of females compared to 44% of males reported practicing physical exercise to

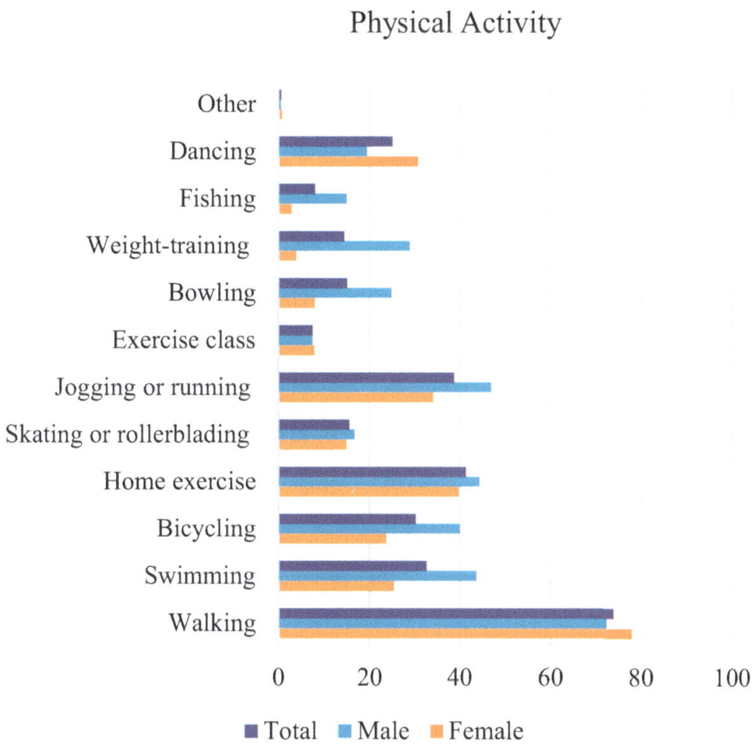


**Fig. 4.1** Prevalence of physical activity among male and female adolescents living in the UAE

maintain health ($p < 0.001$). Forms of PA that were most reported included walking for exercise (74%), home exercise (41%), and jogging or running (39%) (Fig. 1). In addition to those reported in Figs. 1 and 2, participants reported participation in *other* organized sport activities (karate, $n = 13$; hockey, $n = 7$; gymnastics, $n = 4$; boxing, $n = 1$; football, $n = 6$; badminton, $n = 26$, and cricket, $n = 39$), as well as *other* physical activities (skipping, $n = 13$; handball, $n = 2$; yoga, $n = 7$; horseback riding, $n = 25$; treadmill, $n = 2$; wrestling, $n = 1$, and ping pong, $n = 5$). Participation in leisure PA including dancing and walking was significantly higher among girls compared to boys ($p < 0.001$). Interestingly, a higher proportion of girls reported playing basketball than boys (26% versus 22%, $p < 0.01$), while significantly more boys played volleyball, soccer, and golf (Table 4.1). Significant gender differences were found in relation to participation in physical activity, with a higher proportion of boys participating in swimming, bicycling, jogging or running, bowling, weigh training, or fishing, while more girls undertook walking, home exercise, and dancing. Participating in outdoor PA or sport for 7 h or more per week was significantly higher among boys than girls (23% versus 18%, $p < 0.001$). Overall, 3.6% of participants did not participate in any organized sport or PA ($n = 234$). Overall, 56% of participants participated in both organized sport and PA ($n = 3648$). Overall, 11% of adolescents reported less than 2 h of television and computers/video games per day

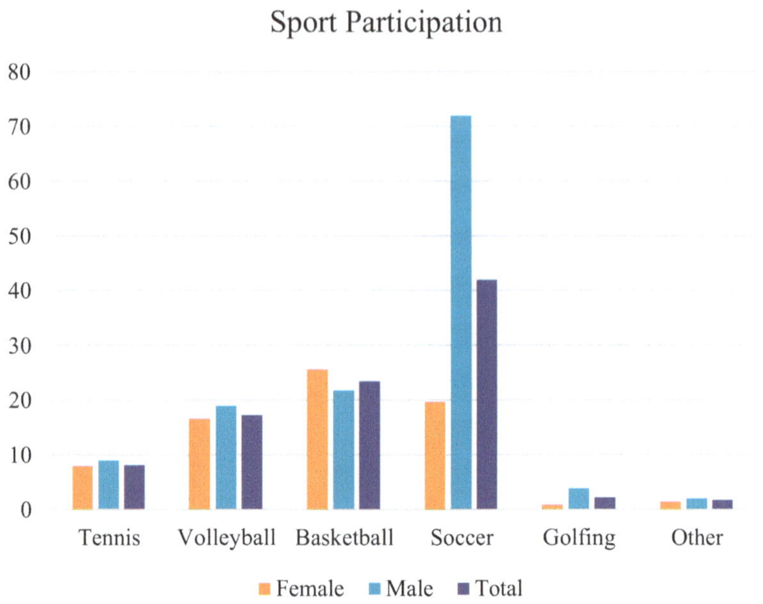


**Fig. 4.2** Prevalence of female and male adolescents who participate in organized sports in the UAE

and thus met the Canadian Sedentary Behavior Guidelines and Global Matrix 3.0 (Aubert et al. 2018).

## 4.4 Discussion

Despite unequivocal evidence on the diverse benefits of PA during adolescence, research highlights high levels of sedentary behavior among adolescents from the MENA, ranging between 65% and 91% (Sarsour et al. 2019; Sharara et al. 2018; Yammine 2017). In addition, studies point to the lack of PA and sport participation among adolescents from MENA. For example, 11–49% of UAE adolescents never participate in PA (Yammine 2017). Gender-specific differences in relation to participation in types of PA or sports and the location (outdoors versus indoors) are apparent among adolescents from MENA. This highlights the importance of PA research that specifically focuses on these factors when assessing the prevalence of physical activity. For instance, research that focuses on aerobic exercises or dance is scarce, although it tends to be quite popular among girls from MENA. In fact, to date, MENA-based studies often examined overall PA without specific attention to different types of physical activity and focused on MET values, which falls short of addressing cultural and social sensitivities (Al-Hazzaa and AlMarzooqi 2018) and may have underestimated the levels of PA among MENA adolescent girls.

**Table 4.1** Prevalence of physical activity and sport participation among male and female adolescents from the UAE ($n = 6356$)

| | Classification | Overall | | Male | | Female | |
|---|---|---|---|---|---|---|---|
| | | *n* | % | *n* | % | *n* | % |
| Sex | | | | 2826 | 45 | 3530 | 56 |
| Practice physical exercise to maintain health*** | Yes | 3595 | 57 | 2000 | 73 | 1540 | 44 |
| Physical activity | Walking*** | 4783 | 74 | 1997 | 72 | 2722 | 78 |
| | Swimming*** | 2118 | 33 | 1196 | 44 | 885 | 25 |
| | Bicycling*** | 1959 | 30 | 1100 | 40 | 821 | 24 |
| | Home exercise** | 2677 | 41 | 1089 | 40 | 1548 | 44 |
| | Skating or rollerblading | 1008 | 16 | 405 | 15 | 586 | 17 |
| | Jogging or running*** | 2509 | 39 | 1289 | 47 | 1188 | 34 |
| | Exercise class | 489 | 8 | 205 | 8 | 274 | 8 |
| | Bowling*** | 974 | 15 | 677 | 25 | 280 | 8 |
| | Weight training*** | 939 | 15 | 793 | 29 | 131 | 4 |
| | Fishing*** | 522 | 8 | 407 | 15 | 100 | 3 |
| | Dancing*** | 1633 | 25 | 531 | 20 | 1074 | 31 |
| | Other | 47 | <1 | 17 | <1 | 30 | <1 |
| Organized sport | Tennis | 530 | 8 | 244 | 9 | 280 | 8 |
| | Volleyball* | 1120 | 17 | 517 | 19 | 581 | 17 |
| | Basketball** | 1518 | 24 | 593 | 22 | 899 | 26 |
| | Soccer*** | 2742 | 42 | 1991 | 72 | 688 | 20 |
| | Golfing*** | 151 | 2 | 106 | 4 | 36 | 1 |
| | Other | 114 | 2 | 59 | 2 | 55 | 2 |
| Outside for 7 or more hours/week*** | Yes | 2678 | 41 | 1509 | 23 | 1132 | 18 |

*$p < 0.05$, **$p < 0.01$, ***$p < 0.001$

In relation to constraints to PA and sport participation, studies have identified unique circumstances specific to the region. A number of studies have attributed low levels of PA to rapid economic development and urbanization that characterizes many countries in this region (Allafi et al. 2014; Henry et al. 2004; Sharara et al. 2018; Wasfi et al. 2008). For example, within the past few decades, rapid economic development in the UAE led to lifestyle changes, such as changes in community design, more passive leisure and entertainment pursuits, and reductions in active transportation, all of which contribute to sedentary lifestyles (Henry et al. 2004; Paulo et al. 2018; Wasfi et al. 2008). Locations or facilities conducive to adolescent physical activity, as well as related opportunities, are noted constraints to PA and sport participation due to these economic and developmental changes. For instance, Wasfi et al. (2008) reported limited indoor spaces available for girls' participation in PA and sport, which is very important given the climate restrictions in the UAE and other MENA countries (Henry et al. 2004). Other studies also identified constraints to PA and sport participation related to limited sport centers and facilities and lack of access to safe parks and playgrounds (Al Sabbah et al. 2007; Fazah et al. 2010). Particularly for girls, accessibility to both indoor and safe outdoor facilities for PA and sport participation has the capacity to increase PA levels and sport participation

(Sharara et al. 2018; Allafi et al. 2014). The availability of outdoor safe space for PA is particularly important and essential, particularly during global crises such as the ongoing COVID-19 pandemic, given that adolescents may not be able to achieve optimal PA levels within the home and equally may not be able to access facilities and sport clubs due to public restrictions. In addition, rapid economic development in many MENA counties led to changes in the occupational interests of the population, as well as changes related to day-to-day house chores. Occupations that require physical exertion generally rely on expatriate labor, as do personal and household duties, such as house cleaning and gardening, which discourage PA and promote sedentary lifestyles (Sharara et al. 2018). Furthermore, economic development led to a heavy reliance on motor vehicles and a general absence of active transport options (Al-Hazzaa et al. 2011).

Social support is another noted constraint to PA and sport participation for adolescents in the MENA. Several studies highlighted the link between adolescent PA and support from friends, parents, and teachers (Bani-Issa et al. 2020; Musaiger et al. 2013; Sharara et al. 2018; Yaffe 2018). Recent studies highlighted that adolescents were more likely to be active when they had the support of their friends and peers (Tesler et al. 2019). Alongside, evidence highlighted parental preferences toward spiritual and educational activities rather than PA and sport participation and focused on how this resulted in low adolescent PA levels and sedentary behaviors (Sharara et al. 2018). Parental education positively related to adolescent PA levels (Al Sabbah et al. 2007; Wasfi et al. 2008). Although evidence of the influence of parental and peer support on adolescent PA level is consistent, some studies have linked this relationship to local geopolitical regional factors. For example, adolescents from the Gaza Strip, particularly boys, may be encouraged and more inclined to give up on PA and sport participation in favor of securing an income and improving their socioeconomic status (Al Sabbah et al. 2007).

While the previously discussed constraints to PA focused on physical and social environmental factors, literature evidence suggests that individual factors play an important role in enhancing adolescent PA levels and sport participation. Literature notes self-efficacy beliefs and self-regulatory skills, which begin to form during adolescence, as constraints to adolescent PA and sport participation (Muturi et al. 2016; Pajares and Urdan 2006). Constraints to PA and sport participation that are above noted, including age, gender, and those related to rapid economic development and social support, interact to contribute to high levels of adolescent sedentary behavior and low levels of physical activity, which in turn influence one's self-efficacy and self-regulatory skills. Thus, the perceived ability of an adolescent to maintain a healthy level of PA may be indirectly correlated to increasing age, gender (F), lack of facilities, and lack of support among friends, peers, and teachers (Al-Hazzaa et al. 2011; Alsubaie and Omer 2015; Musaiger et al. 2013; Sharara et al. 2018). Although findings suggest that self-efficacy in relation to physical activity relates to parental, peer, and teacher influences, and especially opportunities for different forms of exercise during school hours, researchers note that there is limited extant research on the exercise self-efficacy of adolescents in the MENA (Bani-Issa et al. 2020). Furthermore, other individual constraints include hindered individual motivation and interest in PA and sport participation and prioritization of other activities that improve education and social status, largely due to increasing social responsibilities. On the other hand, outcomes of reduced PA and sport participation, such as overweight or obesity, may be elements of a positive feedback loop for decreased levels of PA and sport participation. Fazah et al. (2010) reported normal-weight boys and girls demonstrate higher levels of PA than their obese peers and equally reap the benefits of higher physical functioning and quality of life (Fazah et al. 2010).

The UAE case study suggests that the prevalence of not participating in neither sport nor PA in the last year among adolescents is relatively low (3.6%). While results suggested that over 60% of adolescents regularly participated in some form of PA or sport, study limitations prevented assessments on whether participants met the recommended levels of moderate-vigorous PA levels. In parallel with other MENA-based study findings, results support that practicing PA or exercise to maintain health is significantly higher among boys compared to girls. The type of sport or PA was gender-related, interestingly with girls more likely to play basketball. Indeed, basketball in the MENA has been growing in popularity, making it one of the five top most-participated sports in the region, with women's basketball gaining popularity particularly in Saudi Arabia (Jamali 2014). In addition, girls may be more inclined to play basketball, as it is often an indoor sport and has less physical contact than other competitive sports. As soccer is a traditionally recognized sport in the MENA, the high prevalence of participation in soccer among boys is not surprising. Soccer is often an outdoor sport, and thus its relative unpopularity among girls from the MENA may be due to the cultural and normative construction imposed throughout the MENA for a long period, that only boys practice soccer or "football" in schools. In regard to adolescent preferences to the different type of sports and physical activity, the changes that many MENA countries are experiencing, resulting from globalization, modernization, and educational reform, have provided greater knowledge and awareness of the benefits of PA and sport (Harkness and Hongsermeier 2015).

It is interesting that the sport commercial industry has recently taken note of the gender-related constraints to PA and sport participation. Sporting brands have created sporting attire that appeals to the Muslim girl, as inspired by various Muslim girl athletes around the world, including Fatima Al-Nabhani (Omani tennis player) and Bahraini sprinter Al-Ghasara, as well as more recently Ibtihaj Muhammad, all of whom competed in athletics wearing a hijab (Harkness and Islam 2011; Alvarez 2017). Marketing of sports clothing that is culturally sensitive to MENA populations through big brands such as *Nike*, estimated to be worth five trillion dollars by 2020 (Alvarez 2017), may further increase influence and encourage the breaking of sociocultural norms, in relation to gender-related and cultural factors attached to practicing sport in the MENA. Role model female athletes and culturally sensitive sport attire are key to addressing gender-specific sport constraints. Concurrently, these factors have the capacity to increase girl participation in the MENA, by accounting for hot climate, safety and religious concerns, and cultural trends. All these constraints have been attributed to low levels of PA and sport participation among adolescent girls from the MENA, particularly in relation to outdoor activities and once they reach the age of 12 (Sharara et al. 2018; Musaiger et al. 2013; Obermeyer et al. 2015; Sarsour et al. 2019).

Given the clear social and health benefits associated with being physically active during adolescence, many organizations have emphasized the importance of meeting PA guidelines for adolescents (Tesler et al. 2019). In the MENA, regional constraints to PA and sport participation require appropriately tailored interventions and recommendations. First, there appears to be a need to ensure that physical environments within the MENA region promote adolescent PA and sport participation. This can be achieved through urban design that accounts for sidewalk space, improved walkability, green and park space, as well as outdoor and indoor sport facilities to combat challenges related to climate conditions. An emphasis on the development of active transport systems, walkable environments, and safe spaces may aid in encouraging PA (Paulo et al. 2018). Special attention should focus on opportunities and facilities tailored for adolescent girls' participation in PA and sport. Second, interventions should focus on educating adolescents and their families on the benefits of PA and sport, in order to target related social and cultural perceptions and to emphasize the importance of being physically active (Mohammed et al. 2020). These interventions may be more effective during early years, given evidence that forming early beliefs and perceptions of PA in childhood can develop positive self-efficacy toward an adolescents' perceived ability to maintain exercise self-efficacy and build their skills (Bani-Issa et al. 2020). Lastly, it is essential for future research to further examine constraints to PA among adolescents in the MENA, in order to inform culturally appropriate interventions (Paulo et al. 2018). Studies conducted at various local and regional scales have the capacity to uncover geographically based modifiable constraints. In addition, further research should explore the challenges that girls from the MENA face as these are environmental, cultural, and individual unique constraints that have the capacity to interact and compound the effects of decreased PA and sport participation (Klicnik et al. 2020; Mohammed et al. 2020). By exploring constraints to PA among adolescents, research can inform appropriate interventions to promote PA in the MENA.

## Reflection Questions

- Discuss the benefits of physical activity during adolescence.
- Discuss the role of gender in relation to constraints to physical activity within the MENA region. In your opinion, would this apply to regions outside of MENA?
- How may constraints interact to impact physical activity or sport participation among adolescents in the MENA region?
- Reflecting on sociocultural environments, how can we tackle social disparities to ensure equal opportunities for physical activity among adolescents from the MENA?

## References

Al Sabbah, H., Vereecken, C., Kolsteren, P., Abdeen, Z., & Maes, L. (2007). Food habits and physical activity patterns among Palestinian adolescents: Findings from the national study of Palestinian schoolchildren (HBSC-WBG2004). Public Health Nutrition, 10(7), 739–746. https://doi.org/10.1017/S1368980007665501

Al-Hazzaa, H. M., & AlMarzooqi, M. A. (2018). Descriptive analysis of physical activity initiatives for health promotion in Saudi Arabia. *Frontiers in public health*, *6*, 329. https://doi.org/10.3389/fpubh.2018.00329

Al-Hazzaa, H. M., Abahussain, N. A., Al-Sobayel, H. I., Qahwaji, D. M., & Musaiger, A. O. (2011). Physical activity, sedentary behaviors and dietary habits among Saudi adolescents relative to age, gender and region. International Journal of Behavioral Nutrition and Physical Activity, 8(1), 1–14. https://doi.org/10.1186/1479-5868-8-140

Allafi, A., Al-Haifi, A. R., Al-Fayez, M. A., Al-Athari, B. I., Al-Ajmi, F. A., Al-Hazzaa, H. M., ... & Ahmed, F. (2014). Physical activity, sedentary behaviours and dietary habits among Kuwaiti adolescents: gender differences. *Public health nutrition*, 17(9), 2045–2052. doi:https://doi.org/10.1017/S1368980013002218

Alsubaie, A. S., & Omer, E. O. (2015). Physical activity behavior predictors, reasons and barriers among male adolescents in Riyadh, Saudi Arabia: evidence for obesogenic environment. *International journal of health sciences*, 9(4), 400.

Alvarez. A. (2017). Hijab in Sports: How Muslim Women Athletes Are Fighting for Acceptance. www.rollingstone.com

Aubert, S., Barnes, J. D., Abdeta, C., Abi Nader, P., Adeniyi, A. F., Aguilar-Farias, N., ... & Tremblay, M. S. (2018). Global matrix 3.0 Physical activity report card grades for children and youth: results and analysis from 49 countries. *Journal of Physical Activity and health*, 15(s2), S251–S273.

Bani-Issa, W., Dennis, C. L., Brown, H. K., Ibrahim, A., Almomani, F. M., Walton, L. M., & Al-Shujairi, A. M. (2020). The Influence of Parents and Schools on Adolescents' Perceived Diet and Exercise Self-Efficacy: A School-Based Sample From the United Arab Emirates. *Journal of Transcultural Nursing*, 31(5), 479–491. https://doi.org/10.1177/1043659619876686

Barakat-Haddad C. (2013) Prevalence of high blood pressure, heart disease, thalassemia, sickle-cell anemia, and iron-deficiency anemia among the UAE adolescent population. *Journal of environmental and public health*. https://doi.org/10.1155/2013/680631

Benajiba, N., Mahrous, L., Janah, K., Alqabbani, S. F., Chavarria, E. A., & Aboul-Enein, B. H. (2020). Physical activity interventions promoted in the Arabic-speaking region: A review of the current literature. Obesity Reviews, 21(8). https://doi.org/10.1111/obr.13032

Chaabane, S., Chaabna, K., Abraham, A., Mamtani, R., & Cheema, S. (2020). Physical activity and sedentary behaviour in the Middle East and North Africa: An overview of systematic reviews and meta-analysis. Scientific reports, 10(1), 1–24. https://doi.org/10.1038/s41598-020-66163-x

Craggs, C., Corder, K., Van Sluijs, E. M. F., & Griffin, S. J. (2011). Determinants of change in Physical activity in children and adolescents: A systematic review. *American Journal of Preventive Medicine*, 40(6), 645–658. https://doi.org/10.1016/j.amepre.2011.02.025

Eime, R., Young, J. A., Harvey, J. T., Charity, M. J., & Payne, W. R. (2013). A systematic review of the psychological and social benefits of participation in sport for children and adolescents: Informing development of a conceptual model of health through sport. *International Journal of Behavioral Nutrition and Physical Activity*, 10. https://doi.org/10.1186/1479-5868-10-98

Fazah, A., Jacob, C., Moussa, E., El-Hage, R., Youssef, H., & Delamarche, P. (2010). Activity, inactivity and quality of life among Lebanese adolescents. *Pediatrics International*, 52(4), 573–578. https://doi.org/10.1111/j.1442-200X.2009.03021.x

Hamrani, A., Mehdad, S., El Kari, K., El Hamdouchi, A., El Menchawy, I., Belghiti, H., ... & Aguenaou, H. (2015). Physical activity and dietary habits among Moroccan adolescents. *Public health nutrition*, 18(10), 1793–1800. doi:https://doi.org/10.1017/S1368980014002274

Harkness, G. & Islam, S. (2011). Muslim Female Athletes and the Hijab. Contexts 10(4): 64–65.

Harkness, G., & Hongsermeier, N. (2015). Female sports as non-movement resistance in the Middle East and North Africa. Sociology Compass, 9(12), 1082–1093.

Henry, C. J. K., Lightowler, H. J., & Al-Hourani, H. M. (2004). Physical activity and levels of inactivity in adolescent females ages 11-16 years in the United Arab Emirates. *American Journal of Human Biology*, 16(3), 346–353. https://doi.org/10.1002/ajhb.20022

Jamali, H. (2014, October 29). Women in Saudi Arabia Use Basketball to Push for More Rights (Rep.). Retrieved from https://www.nbcnews.com/news/world/women-saudi-arabia-use-basketball-push-more-rights-n236336

Kaluski, D. N., Mazengia, G. D., Shimony, T., Goldsmith, R., & Berry, E. M. (2009). Prevalence and determinants of physical activity and lifestyle in relation to obesity among schoolchildren in Israel. Public health nutrition, 12(6), 774–782. doi:https://doi.org/10.1017/S1368980008002991

Klicnik, I., Mohammed, S., & Barakat, C. (2020). Development and Validation of a Survey Instrument to Build Capacity for Examining Constraints to Sport Participation: A Pilot Study. Manuscript in Preparation.

Mohammed, S., Kerpan, S., Stanyon, W., & Barakat, C. (2020). Examining Constraints to Sport Participation among Immigrant Adolescent Girls who reside in the Greater Toronto Area. *Archives of Health Science*, 4(1), 1–13.

Mohammed, S., Kerpan, S., Stanyon, W., & Barakat, C. (2021). Exploring Experiences of Immigrant Adolescent Girls on Constraints to Sport Participation. Manuscript in Preparation.

Musaiger, A. O., Al-Mannai, M., Tayyem, R., Al-Lalla, O., Ali, E. Y. A., Kalam, F., … Chirane, M. (2013). Perceived barriers to healthy eating and physical activity among adolescents in seven Arab countries: a cross-cultural study. Scientific World Journal, 232164. https://doi.org/10.1155/2013/232164

Muturi, N. W., Kidd, T., Khan, T., Kattelmann, K., Zies, S., Lindshield, E., & Adhikari, K. (2016). An examination of factors associated with self-efficacy for food choice and healthy eating among low-income adolescents in three U.S. states. Frontiers in Communication, 1, 6. doi:https://doi.org/10.3389/fcomm.2016.00006

Obermeyer, C. M., Bott, S., & Sassine, A. J. (2015). Arab Adolescents: Health, Gender, and Social Context. *Journal of Adolescent Health*, 57(3), 252–262. https://doi.org/10.1016/j.jadohealth.2015.01.002

Pajares, F., & Urdan, T. (2006). Adolescence and education: Self efficacy beliefs of adolescents (Vol. 5). Greenwich, CT: Information Age

Paulo, M. S., Nauman, J., Abdulle, A., Aljunaibi, A., Alzaabi, M., Barakat-Haddad, C., … Loney, T. (2018). Results from the United Arab Emirates' 2018 Report Card on Physical Activity for Children and Youth. Global Health Action, 15(1), 419–422. https://doi.org/10.3402/gha.v6i0.20100

Ridley, K., Ainsworth, B.E. & Olds, T.S. (2008). Development of a Compendium of Energy Expenditures for Youth. Int J Behav Nutr Phys Act 5, 45. https://doi.org/10.1186/1479-5868-5-45

Sarsour, A. K., Turban, M., Al Wahaidi, A., Abed, Y., & Alkahlout, H. (2019). Does gender influence food intake and PApattern among palestinian adolescents in the gaza strip? Eastern Mediterranean Health Journal, 25(10), 722–727. https://doi.org/10.26719/emhj.19.073

Sharara, E., Akik, C., Ghattas, H., & Obermeyer, C. M. (2018). Physical inactivity, gender and culture in Arab countries: a systematic assessment of the literature. BMC Public Health, 18(1), N.PAG-N.PAG. https://doi.org/10.1186/s12889-018-5472-z

Tesler, R., Kolobov, T., Ng, K. W., Shapiro, E., Walsh, S. D., Shuval, K., & Harel-Fisch, Y. (2019). Ethnic disparities in physical activity among adolescents in Israel. *American Journal of Health Behavior*, 43(2), 337–348. https://doi.org/10.5993/AJHB.43.2.10

Tremblay, M. S., LeBlanc, A. G., Janssen, I., Kho, M. E., Hicks, A., Murumets, K., ... & Duggan, M. (2011). Canadian sedentary behaviour guidelines for children and youth. *Applied Physiology, Nutrition, and Metabolism*, *36*(1), 59–64. https://doi.org/10.1139/H11-012

Wasfi, A., El-Sherbiny, A., Gurashi, E., & Al Sayegh, F. (2008). Sport Practice Among Private Secondary Students in Dubai in 2004. La Revue de Sante de La Mediterranee Orientale, 4(3).
WHO (2020). https://www.who.int/news-room/fact-sheets/detail/physical-activity
Yaffe, Y. (2018). Physical Activity Among Israeli-Arab Adolescent Males: How Do Parenting Styles Matter? American Journal of Men's Health, 12(6), 2037–2043. https://doi.org/10.1177/1557988318790881
Yammine, K. (2017). The prevalence of physical activity among the young population of UAE: A meta-analysis. Perspectives in Public Health, 137(5), 275–280. https://doi.org/10.1177/1757913916675388
Zaabi MA, Shah SM, Sheek-Hussein M, et al. (2016). Results from the United Arab Emirates' 2016 report card on physical activity for children and youth. J Phys Act Health, 13(11 suppl 2):S299–S306. doi:https://doi.org/10.1123/jpah.2016-0312

# Chapter 5
# Predictive Modeling for Obesity and Overweight in Adolescents, Current Status and Application to the MENA Region

**Kinda Khalaf, Dhanya Menoth Mohan, Nour El Asswad, and Fatme Al Anouti**

## Key Highlights

- Context
  - Adolescent obesity is highly prevalent in the Middle East and North Africa and has been mostly attributed to the rapid urbanization and the sudden transition in nutrition and lifestyle.
  - Predictive models for adolescent obesity are important for prevention and are used to calculate the risk of an adolescent of becoming overweight or obese in the future.
- Objectives
  - To highlight the multifactorial aspects of adolescent and childhood obesity.

K. Khalaf (✉) · D. M. Mohan
Department of Biomedical Engineering, Khalifa University, Abu Dhabi, United Arab Emirates
e-mail: kinda.khalaf@ku.ac.ae; dhanya.mohan@ku.ac.ae

N. El Asswad
Department of Biology, University of Toronto, Toronto, ON, Canada
e-mail: nour.elasswad@mail.utoronto.ca

F. Al Anouti (✉)
College of Natural and Health Sciences, Zayed University, Abu Dhabi, United Arab Emirates
e-mail: Fatme.AlAnouti@zu.ac.ae



C. Barakat et al. (eds.), *Adolescent Health in the Middle East and North Africa*, Global Perspectives on Health Geography,
https://doi.org/10.1007/978-3-030-92107-1_5

  - To propose a suitable predictive tool for estimating overweight/obesity among adolescents specifically within the MENA.
- Methodology
  - Several databases were used to retrieve literature about adolescent obesity and predictive algorithms/models between the years 2015 and 2020 in MENA adolescent populations.
  - A list of potential predictors and risk factors for suitable future implementation and validation were identified and used to propose a predictive model.
- Key Findings
  - There is a lack of an accurate population-based predictive algorithms for obesity for this target group.
  - Based on literature review, it has been identified that sociodemographic factors, physical activity, diet, screen time, parental obesity, family history of obesity, and duration of sleep are the important risk factors applicable to the MENA population.
- Conclusion and Implications
  - Based on a holistic approach of integrating relevant main risk factors, we proposed an obesity predictive model that could be valuable for informing policy-makers in the region and helping them set up effective national and regional surveillance systems.
  - Future work includes exploring machine learning-based powerful prediction algorithms for adolescent obesity and overweight by acquiring and incorporating a population-specific vast amount of clinical data.
  - The need to examine the various predictive algorithms in the context of different obesity indicators, in addition to BMI, remains an important aspect for establishing a valid predictive model for adolescent obesity.

## 5.1 Introduction

### *5.1.1 Prevalence of Overweight and Obesity*

With more than two billion people (30% of the world's population) overweight or obese and an annual cost of $2.1 trillion, obesity poses severe global health and economic (World Health Organization 2020a). Comprising the fifth leading risk factor for mortality in the world (around 3.4 million annual deaths), obesity significantly increases the risk for developing numerous chronic diseases, including coronary heart disease (by over 50%), ischemic stroke (by 44%), type 2 diabetes (by 23%), as well as many cancers (up to 41%) (World Health Organization 2020a). According to the McKinsey Global Institute (MGI), a devastating $2.1 trillion, or

2.8% of global GDP, is spent on obesity-related health problems annually but less than 1% on prevention (McKinsey Global Report 2015). The problem is expected to worsen, where almost half of the world's adult population is expected to be overweight or obese by 2030 (McKinsey Global Report 2015). The global trend of sustained growth in obesity prevalence indicates that the current measures in the prevention, treatment, and management of the condition are largely ineffective.

The World Health Organization (WHO) 2020 data reveals that the UAE currently ranks fifth in the world in obesity at a prevalence rate of 36% (33% males and 39% females). Three in every ten Emirati males and almost four out of every ten females are obese, with an economic burden amounting to $6 billion/year in associated disease cost (McKinsey Global Report 2015). If we also include the percentage of overweight individuals, based on the most recent Global Burden of Disease report (World Health Organization 2020a), more than 60% of men and 66% of women in the UAE are currently overweight or obese (average of 63% or more than double the global average of 30%). Furthermore, while the UAE slightly fares better than the USA in adult obesity prevalence (US current rate is 38%), Emirati children are 1.8 times more obese than their American counterparts. This reveals a dangerous future trend and prognosis, particularly considering the very young median age of the population (30.3 years), since obesity is an independent risk factor for both T2D and CVD, the major culprits for mortality and morbidity in the UAE (Health Authority Abu Dhabi 2016).

### *5.1.2 Overweight and Obesity Among Adolescents*

In the last couple of decades, the prevalence of obesity has substantially increased during adolescence, a unique stage of human development and a critical time for maintaining good health while experiencing rapid physical, cognitive, and mental growth (World Health Organization 2020a). In 2016, over one in six adolescents aged 10–19 years, worldwide, were overweight (World Health Organization 2018). The WHO defines adolescence as the phase of life between childhood and adulthood, from ages 10 to 19 years old. According to the WHO global data, the prevalence of overweight and obesity among children and adolescents has dramatically risen from 4% in 1975 to approximately 18%, with a total of 340 million obese adolescents in 2016 (World Health Organization 2020a). Importantly, obesity-related diseases are also increasing sharply among adolescents, where three-quarters of all deaths by the year 2020 are expected to be due to non-communicable diseases (World Health Organization 2020a).

In the Middle East region, especially the Gulf Cooperation Council (GCC) countries (UAE, Saudi Arabia, Kuwait, Bahrain, Oman, Qatar), the prevalence of overweight and obesity among adolescents is considered among the highest in the world (Farrag et al. 2017). Musaiger et al. (2012) reported a relatively high percentage of overweight among adolescents (15–18 years) in all studied countries (Algeria, Jordan, Kuwait, Libya, Palestine, Syria, and the United Arab Emirates), ranging from 9.3% in Algeria to 25.6% in Kuwait. Furthermore, several studies indicate that while the prevalence of childhood obesity in the Middle East and GCC is higher among males (Farrag et al. 2017; Musaiger et al. 2013), obesity prevalence among females is quite high when compared to other countries (Al Hammadi and Reilly 2019). The high prevalence of adolescence obesity in the Middle East and GCC has been mostly attributed to the rapid urbanization, which has led to sudden transition in nutrition and lifestyle (Musaiger et al. 2013). A relevant example is the United Arab Emirates (UAE), where the prevalence of obesity and overweight in adolescence has increased by two- to threefold from 1975 to 2017 (Baniissa et al. 2020; AlBlooshi et al. 2016). Al Hammadi and Reilly (2019) used the WHO definition of obesity and identified more than one-third of the sample in the secondary school-age participants as obese or overweight, indicating an increase in prevalence associated with increasing age. According to AlBlooshi et al. (2016), obesity was investigated in the UAE using different methods for BMI interpretation with different outcomes for the same population (6–17 years old), and hence obesity trends in the UAE among adolescents remain unclear.

### *5.1.3 BMI-Based Obesity Assessment: Implications and Its Limitations*

The main tool currently used worldwide for the assessment of obesity is the body mass index or BMI. Defined as the ratio of the weight of an individual (kg) divided by the square of their height ($m^2$), a BMI ratio of >25 indicates overweight, while that of >30 defines obesity (Centers for Disease Control 2020). Developed in the mid-1800s, there are some issues regarding BMI's validity for obesity assessment and associated chronic disease prediction (Centers for Disease Control 2020; National Institute of Health). For example, BMI does not take age, sex, bone structure, fat distribution, or muscle mass into consideration, all potentially important factors in obesity disease prediction. In general, there are three main sources of error when using BMI: (1) BMI is an indirect measure of obesity, (2) errors in self-reported data, and (3) the poor sensitivity and specificity of BMI. Moreover, there is now strong evidence that the cutoffs of BMI, provided by the WHO, do not adequately reflect the overweight or obesity status of all ethnic populations (ACOG committee opinion 2017). For example, a higher body fat percentage is correlated with lower BMIs among Asians, while among Pacific Islanders, higher BMIs are associated with more muscle mass and less body fat (Wang et al. 2006). Most importantly, BMI has been shown as an unreliable predictor of the risk of chronic diseases associated with obesity, such as cardiovascular and type 2 diabetes (T2D) (ACOG committee opinion 2017). Disagreement on the optimal cutoffs linking BMI to disease risk is cited as one of the main reasons behind ineffective early disease prediction and subsequent intervention (Akhabue et al. 2018).

Based on BMI, obesity is divided into three classes: class 1, BMI ranges from 30 to <35; class 2, BMI ranges from 35 to <40; whereas, class 3 is considered as extreme or severe obesity with BMI at 40 or higher (Centers for Disease Control 2020; National Institute of Health). However, adopting one standard universal concept for obesity is not sufficient.

### 5.1.4 Risk Factors of Adolescent Obesity

A multifactorial condition, obesity is the result of an intricate interplay between genetic, environmental, and inflammation factors with severe functional implications on the neuromusculoskeletal system. Indeed, the literature is rich with many of these factors, including genetics, ethnicity, hormonal and metabolic disorders, adipose tissue distribution imbalances and inflammation, pathogens including viruses and microbiomes, mental stress, sleeping disorders, the effects of high income and urbanization, as well as built environment characteristics including fast food restaurants, available transportation, and lack of walkability (Hruby and Hu 2015). Additional risk factors, which can be detected as early as infancy and early childhood, include high birth weight, rapid weight gain, little or no breastfeeding, early introduction of solid food, maternal and paternal obesity, and maternal smoking during pregnancy (Redsell et al. 2016).

### 5.1.5 Obesity Prediction Models: Relevance and Application

Despite the work on many aspects of obesity, remarkably few studies have addressed its multifactorial nature, while fewer devised quantitative predictive measures. Predictive measures of obesity are key to prevention, which is a good approach to tackle the burden of obesity since it is difficult to reverse once established. Predictive models for adolescents are used to calculate an adolescent's risk of becoming overweight or obese in the future (Butler et al. 2018). Models that have good sensitivity and specificity should differentiate between high-risk individuals and low-risk individuals. Sensitivity is defined as a model's ability to correctly predict (within an acceptable error) individuals who have or may develop obesity. On the other hand, specificity refers to a model's ability to correctly predict the individuals who do not have or will not develop the condition (Butler et al. 2018). Rautiainen and Äyrämö (2019) analyzed studies which used different types of predictive models of overweight/obesity between the ages of 2 and 33 years, including logistic regression, decision trees, Bayesian, and neural networks and support vector machines. A number of features were used to predict obesity at a certain age, such as gender, height, weight, maternal pre-pregnancy weight status, paternal BMI, maternal smoking during pregnancy, breastfeeding during the first year, number of household members, etc. (Rautiainen and Äyrämö 2019). Several predictive models are available for different populations. However, to the best of the authors' knowledge, there are no published work to date that address overweight/obesity predictive algorithms for the

MENA region, including the GCC in spite of the high prevalence and early onset of obesity and associated non-communicable disease in the region. The main objective of this chapter is to review the available obesity models and algorithms toward proposing a suitable predictive tool for estimating overweight/obesity among adolescents, specifically within the MENA. In particular, this chapter highlights the following:

1. A review of various predictive modeling algorithms applicable for overweight/obesity prediction among adolescents
2. Obesity risk factors relevant to the MENA region
3. A list of potential predictive modeling approaches for future implementation and validation for the MENA region

## 5.2 Methodology

### *5.2.1 Search Strategy*

A literature search was conducted within the databases of Google Scholar, Scopus, Web of Science, and PubMed using the following terms: (1) "obesity" OR "overweight" OR "excess weight" OR "BMI" OR "body mass index," (2) "childhood" OR "adolescent" OR "child," (3) "risk factors" OR "lifestyle" OR "socioeconomic" OR "diet" OR "predictors" OR "nutrition" OR "sociodemographic" OR "physical activity" OR "health," (3) "Region" ("Algeria" OR "Bahrain" OR "Egypt" OR "Iran" OR "Kuwait" OR "Lebanon" OR "Morocco" OR "Oman" OR "Palestine" OR "Qatar" OR "Saudi Arabia" OR "Syria" OR "Tunisia" OR "Turkey" OR "United Arab Emirates" OR "Yemen" OR "Middle East and North Africa" OR "MENA" OR "gulf"), and (4) "prediction model" OR "data mining" OR "predictive" OR "algorithm" OR "machine learning" OR "big data." The articles were first screened for title, followed by abstract when it was difficult to include/exclude the same based on the title. All articles published in English within the last 20 years (2000–2020) which met the inclusion criteria were selected. The full text of the published articles was retrieved and further studied.

### *5.2.2 Inclusion and Exclusion Criteria*

Papers were included if they addressed (1) the design and implementation of predictive algorithms or prediction tools aimed at predicting adolescent overweight/obesity, (2) the risk factors associated with overweight/obesity among adolescents in the MENA region, and (3) sample population of adolescents aged 10–19 years. Studies involving subjects outside the age limit, as well as those investigating the outcome of subjects receiving special medication, were outside the scope of this review.

### 5.2.3 Data Extraction

The articles were reviewed to understand the different algorithms available for predicting adolescent obesity and/or overweight. Tools that have been implemented for different cohorts, including the MENA were hand-searched. The characteristics of the articles, including author, year, country, study population characteristics in terms of age, BMI, and outcome measures, are tabulated (Table 5.1).

For a particular model, identification of a set of predictors, for example, in this context, the risk factors associated with overweight/obesity, is key to its success. With the aim to propose a suitable prediction model for the MENA, an additional search was conducted to identify articles that estimated various risk factors associated with overweight and obesity among adolescents in the MENA region (see Table 5.2). Sixteen countries/territories which constitute the vast majority of this region were considered in this review. These regional studies have identified a number of independent variables, including anthropometric, sociodemographic, socioeconomic, as well as clinical factors as well as association with the dependent variable, i.e., obesity.

In summary, the purpose of the search was to identify the most relevant predictive algorithms for adolescent overweight/obesity, as well as to propose a list of potential predictors/risk factors particular to the MENA region for proper future implementation and validation.

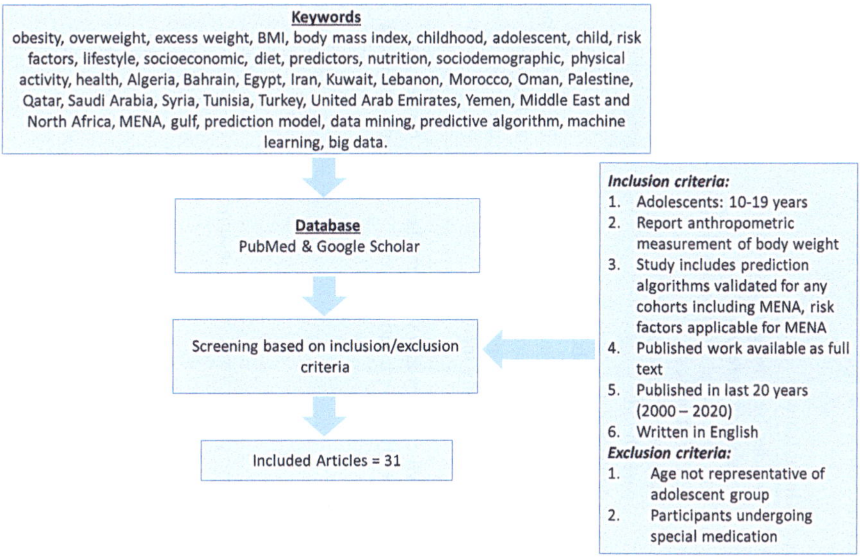


**Fig. 5.1** Flow diagram of the search strategy

**Table 5.1** Literature on overweight/obesity predictive algorithms applicable for adolescents

| Study and dataset/cohort | Input features | Input data recorded at | Prediction model | Prediction outcome | Prediction age | Model performance |
|---|---|---|---|---|---|---|
| Morandi et al. (2012), Finland | Gender, birth weight, gestational weight gain, pre-pregnancy and gestational smoking of mother, BMI and professional category of parents, single parenthood, number of household members | Newborn and parental data | Stepwise logistic regression | Overweight and obesity | Childhood (7 years), adolescents (16 years) for sample development | Adolescent obesity: AUC = 0.75 [0.71–0.79], $p < 0.001$ Adolescent overweight/ obesity: AUC = 0.71 [0.69–0.73], $p < 0.001$ |
| Pei et al. (2013), German | Birth weight, standardized BMI at the age of 5, education status of parents, gestational smoking of mother, family income | Birth to 5 years and parental data | Logistic regression models | Overweight | 10 years | Mean values of: Sensitivity = 37.1% Specificity = 96.5% PPV = 72.5% NPV = 86.1% |
| Riedel et al. (2014), German | BMI category at 6 years, mother's obesity, and education level | 6 years and parental data | Classification and regression tree | Overweight/ obesity | 14 years | BMI ≥ P75 at the age of 6 explained 63.5% [95%CI: 51.1;74.5] and 72.0% [95%CI: 60.4;81.8] of overweight/ obesity at the age of 14 in boys and girls, respectively |
| Jator (2014), existing Medical Expenditure Panel Survey (MEPS) data | Age, gender, race, family income | 12–17 years | Modified logistic model, concentration index | Degree of obesity inequality | 12–17 years | n/a |

| Study and dataset/cohort | Input features | Input data recorded at | Prediction model | Prediction outcome | Prediction age | Model performance |
|---|---|---|---|---|---|---|
| Graversen et al. (2015), Finland | Birth weight, maternal BMI, childhood BMI | Birth, 5 years, 8 years | Logistic regression | Overweight including obesity | 13–16 years | Sensitivity:<br>Birth (F/M): 24.0/ 17.4%<br>5 years: 38.9/28.2%<br>8 years: 49.2/38.7%<br>Specificity:<br>Birth: 92.1/91.7%<br>5 years: 94.4/94.2%<br>8 years: 96.0/96.7% |
| Hudda et al. (2019), UK | Weight, height, age, sex, ethnic group | 4–15 years (development dataset), 11–12 years (external validation dataset) | Multivariate linear regression | Fat mass | 4–15 years, 11–12 years (external validation) | Optimism adjusted $R^2$: 94.8%, 95% confidence interval 94.4% to 95.2%<br>External validation: $R^2$:<br>90.0%, 95% confidence interval 87.2–92.8% |
| Kim et al. (2019), Korean | Region, academic_performance, pressure, suicide_thought, sleeping_quality, drinking, smoking, education_father, education_motherwealth, pocket_money, healthy eating, unhealthy_eating, exercise_60 min, exercise_20 min, sitting_time_study(min), smartphone_time(min), smartphone_service, obesity_level | Adolescents (12–18) | General Bayesian network embedded with Markov blanket | Obesity | Adolescents (12–18) | Accuracy: 53.703%, F-measure: 0.535, and AUC: 0.758 |

**Table 5.2** A summary of literature addressing risk factors of adolescent overweight/obesity in MENA

| Country | Study | Criteria | Age | Risk factors |
|---|---|---|---|---|
| Algeria | Allioua et al. (2015) | IOTF | 10–17 | High calories and unbalanced diets, reduced physical activity |
| Bahrain | Musaiger et al. (2014) | Percentiles of National Health and Nutrition ExaminationSurvey-1 (NHANES-1) growth standard | 15–18 | Significant factors include:<br>Mother's education (higher education with obese children) for both male and female<br>Father's education, rank among siblings, burger size and French fries portion, watching TV >3 h/day (males).<br>Protective factors:<br>1. Eating during school breaks, bringing food from home (for female obesity)<br>2. Eating breakfast at home, eating in-between breakfast and lunch, eating between lunch and dinner (for male obesity).<br>Not significant:<br>Intake of vegetables, dairy products, meat, fish, chicken, legumes, canned juices, and chocolates<br>Fast food intake and soft drink size (both male and female)<br>Frequency of consumption of soft drinks and sweets was negatively associated with obesity in males<br>Fruit intake more than thrice a week reduced the risk in males. |
| Egypt | El-Gilany and El-Masry (2011) | CDC | 14–19 | Risk factors include eating starchy food thrice or more per week<br>1. Physical inactivity<br>2. Fast food/snacking/sweets intake ≥3 times/week<br>3. Television viewing ≥ 2 h/day<br>4. Positive family history of obesity<br>Consumption of vegetables/fruits, plant protein, and dairy products ≥3 time/week was protective factor |
| | Talat and El Shahat (2016) | n/a | 12–15 | Low parent education level, skipping breakfast, eating snacks as a substitute, fast food intake, physical activity decline, TV viewing duration >2 h/day, snacking during TV watching |

(continued)

**Table 5.2** (continued)

| Country | Study | Criteria | Age | Risk factors |
|---|---|---|---|---|
| Iran | Abiri et al. (2019) | CDC | 14–17 | Physical activity, computer use, duration of breastfeeding, total sleep time, parental education, economic status |
| Kuwait | Al-Haifi et al. (2013) | IOTF (for 14–17)<br>WHO(for ≥18) | 14–19 | Associated factors are:<br>Physical activity<br>Eating habit: consumption of breakfast (both boys and girls), vegetables (only boys), and fast foods (boys and girls) and potatoes, cakes and doughnuts, and sweets (girls only)<br>No association with:<br>1. Sedentary behaviors, TV viewing, computer usage |
| Lebanon | Nasreddine et al. (2014) | BMI z-score according to the WHO new growth standard.<br>IOTF, CDC | 6–19<br>(12–19 adolescents) | Positive correlation with male gender, mother's employment status, residence in Beirut, sedentary time, eating fast food, and sugar-sweetened beverages<br>Negative association:<br>1. Borderline significant association between higher physical activity and lower odds of overweight<br>2. Increased intakes of milk/dairy products<br>3. Regular breakfast consumption |
| Morocco (Fez) | El Kabbaoui et al. (2018) | WHO 2007 | 12–18 | Higher education of father or mother, higher family income, motorized transport to school, computer usage >4 h/day, frequent intake of soda and soft drinks<br>No association between overweight/obesity and sleep, TV screen time, physical activity |
| | Nouayti et al. (2020) | IOTF | | Urban residence, father's income (≥ 5000 MAD), and overweight/obesity, female sex |
| Oman | Waly et al. (2017) | – | 17.2 ± 1.4 | Sedentary lifestyle, unhealthy nutritional habits |
| Palestine | Jildeh et al. (2011) | CDC, IOTF | 11–16 | Less physical activity, lower/inadequate energy intake |
| | Mikki et al. (2009) | CDC, IOTF | 13–15 | High standard of living among boys, onset of puberty among girls |

(continued)

**Table 5.2** (continued)

| Country | Study | Criteria | Age | Risk factors |
|---|---|---|---|---|
| Qatar | Kerkadi et al. (2019) | IOTF | 14–18 | Significant factors included being male and skipping breakfast.<br>No association between obesity (general and abdominal) and screen time, physical activity<br>Negative association between obesity and intake of unhealthy foods. |
| | Bener et al. (2011) | Qatari growth pattern curves | 6–18 | Watching television for more than 4 h, lack of sleep (5–7 h or less) |
| Saudi Arabia | Al-Hazzaa et al. (2012) | IOTF | 14–19 | Less frequency of vigorous physical activity (in both male and female), skipping breakfast or infrequent consumption of vegetables, frequent consumption of sugar-sweetened beverages |
| | Amin et al. (2008) | According to Cole et al. (2000) | 10–14 males | Urban residence, older age of children, mother's low education status, mother's occupational status, and family size ≤6, consumption of food away from home, infrequent breakfast intake at home, frequent consumption of sweets/candy and carbonated drinks, low servings of vegetables, fruits, and dairy products |
| Syria | Nasreddine et al. (2010) | WHO 2007 WHO 1995 and IOTF for comparison | 15–18 | Male gender, positive family history of obesity, increased educational attainment for both parents, lower crowding index than their counterparts, energy consumption from carbohydrate |
| Tunisia | Aounallah-Skhiri et al. (2008) | WHO<br>IOTF for comparison purpose | 15–19 | Males: (Rural area) – working mother, low physical activity<br>(Urban area) – irregular snacking<br>Female: (Rural) – mother's education level<br>(Urban) – not attending school, skipping daily meals |
| | Zarrouk et al. (2009) | Cole et al. (2000) | 8–11 | Physical activity, sedentary time |
| Turkey | Pirinçci et al. (2010) | IOTF | 6–11 | Snacking during television watching, fast food intake |
| | Discigil et al. (2009) | CDC 2000 | 6–16 | High socioeconomic status, preschool care source other than mother<br>No association with gender, adolescence, parental education level, and occupational status of father |

(continued)

**Table 5.2** (continued)

| Country | Study | Criteria | Age | Risk factors |
|---|---|---|---|---|
| United Arab Emirates | Al Junaibi et al. (2013) | CDC | 6–19 | Risk factors include older age, male gender, lack of dairy consumption, higher parental BMI<br>No association with physical activity or family income |
| | Kerkadi et al. (2005) | CDC | 5–14 | Daily intake of breakfast, television watching time more than 2 h/day, parental obesity, physical activity |
| Yemen | Raja'a and Mohanna (2005) | | 10–18 | Private schooling, higher in females, sedentary life style, family history of obesity, education level of father, consumption of unhealthy foods |

## 5.3 Results/Findings

### *5.3.1 Predictive Modeling/Algorithm: An Overview*

With the advent of faster computers, artificial intelligence, and big data tools and methodologies, the concept of predictive modeling has made great strides in healthcare over the past few years. Sophisticated models with accepted levels of accuracy enable healthcare practitioners to devise personalized treatment strategies, as well as early preventive measures in a cost-effective manner. These models use machine learning-based techniques and/or statistical approaches to infer trends in data and predict a future event in real time.

In general, the design and implementation of predictive modeling include four important phases: (i) problem definition and data collection, (ii) model development and internal validation, (iii) model testing in a real-world setting, and (iv) broader dissemination (Cohen et al. 2014). During the initial phase, the problem is defined, and patient data are acquired through electronic health records or other means. A model is then generated using various mathematical relationships, ranging from simple regression methods to complex artificial intelligence-based approaches, such as support vector machines and neural networks. A combination of clinical data is used as input predictors. Further, the model with the highest prediction accuracy is tested and validated on the internal dataset and subsequently on real-world settings. Model validation is considered to be an important phase as it helps the research community to quantify the predictive validity of a model. It explains the applicability of a model derived using one dataset on a completely new dataset (Ivanescu et al. 2016).

### 5.3.2 *Predictive Modeling Approaches for Adolescent Overweight or Obesity*

The literature search conducted here identified several prediction tools/algorithms aimed at predicting the occurrence of overweight/obesity in adolescents, considering birth, parental, and/or early childhood data as input features (Table 1). To the best of the authors' knowledge, there are no published works to date that address overweight/obesity predictive algorithms for the MENA region.

Morandi et al. applied stepwise logistic regression analysis on a Northern Finland Birth Cohort and claimed that adolescent obesity could be predicted at birth, with maternal BMI as the strongest predictive candidate and the genetic score as the most modest with regard to the prediction accuracy. This study reported an AUC of 0.75 and 0.71 for adolescent obesity and adolescent overweight/obesity, respectively. In a similar study cohort, Graversen et al. (2015) designed a logistic regression model based on birth weight, maternal BMI, and childhood BMI to predict adolescent overweight and adult overweight/obesity. Internal validation was conducted using the Northern Finland Birth Cohort born in 1966, and external validation was performed on the Northern Finland Birth Cohort born 20 years later, where the prevalence of overweight was high, resulting in satisfactory to good prediction outcomes. Another study developed a similar model using a set of risk factors, including birth weight, standardized BMI at the age of 5 years (60–64 months), parental education, family income, and maternal gestational smoking (Pei et al. 2013). High BMI/overweight at 5 years was found to be a strong predictor of being overweight at 10 years. Although the sensitivity (37.1% for the combined sample) of this model was low, the specificity reached up to 96.5% for the combined data.

Riedel et al. (2014) adopted a classification tree approach to predict overweight/obesity at the age of 14 from BMI calculated at 6 years of age, as well as the education level and obesity of the mother. The BMI value at age 6 was an important predictor in agreement with (Pei et al. 2013).

A modified logistic regression model was used in (Jator 2014) to measure obesity distribution among different races, considering age, gender, race, and family income as input predictors. This study revealed a negative association of family income with adolescent obesity. On the other hand, Hudda et al. developed a multivariate model to predict fat mass levels in children aged 4–15 years (external validation in children aged 11–12 years), where input predictors included simple anthropometric and demographic variables (height, weight, age, gender, and ethnicity) (Hudda et al. 2019). This model showed excellent predictive performance (optimism adjusted $R^2$: 94.8%) on the internal validation promising generalizability. Kim et al. utilized a general Bayesian network embedded with Markov blanket combined with a what-if analysis used for adolescent obesity prediction (Kim et al. 2019). Several parameters as reported in Table 5.1 are considered.

### 5.3.3 *Predictive Variables/Predictors Applicable for the MENA Region*

Identifying a set of significant predictive variables or predictors that are strongly correlated to the prediction outcome is crucial during model development. Table 2 highlights research articles that report various risk factors associated with adolescent overweight/obesity in the MENA region.

In general, the risk factors of obesity can broadly be categorized as sociodemographic, dietary, and lifestyle factors. A vast amount of literature is available addressing the association between various risk factors and overweight/obesity.

**Sociodemographic Factors**

The association between overweight/obesity and sociodemographic factors was highlighted by several research studies (Talat and El Shahat 2016; Musaiger et al. 2014; Amin et al. 2008). For example, it has been identified that the mother's higher educational level is linked to obesity in both male and female adolescents in Bahrain, whereas the father's educational level and rank among siblings were significant only among males (Musaiger et al. 2014). These findings contradict the research outcome from Egypt (Talat and El Shahat 2016) and Saudi Arabia (Amin et al. 2008), where low parental education was positively associated with obesity. The effect of parental education was also reported from Iran (Abiri et al. 2019), Tunisia (Aounallah-Skhiri et al. 2008), and Yemen (Raja'a and Mohanna 2005). Besides, no association was found between obesity and parent's education in a study conducted in Turkey (Discigil et al. 2009).

In Morocco, based on logistic regression analysis, a study (El Kabbaoui et al. 2018) found that higher education of parents, higher family income, and motorized transport to school were significant risk factors of excess weight. In line with this study, most recently, Nouayti et al. (2020) reported that father's income higher than 5000 Moroccan Dirhams and urban residence were risk factors among Moroccan adolescents. The higher standard of living index was associated with obesity among adolescent males in Palestine (Mikki et al. 2009). Another study suggested that socioeconomic factors, including the educational attainment of parents as well as lower crowding index (the number of members in a household divided by the total number of rooms, excluding kitchen and bathrooms), were strong predictors (Nasreddine et al. 2010). In Lebanon, urban residence and maternal employment contribute to higher risk of obesity. This corroborates the findings from Saudi that revealed the significance of urban residence and mother's occupational status, in addition to family size (less than 6) (Amin et al. 2008). However, Al Junaibi et al. (2013) found no association between obesity and family income in the UAE population (Al Junaibi et al. 2013).

Nasreddine et al. highlighted that prevalence is higher among male adolescents in Lebanon (Nasreddine et al. 2014) and Syria (Nasreddine et al. 2010). Similar findings were reported by Kerkadi et al. (2019) and Al Junaibi et al. (2013) based on multivariate analyses. However, it was found that female adolescents in Morocco (Nouayti et al. 2020) and Yemen (Raja'a and Mohanna 2005) were at increased risk of developing obesity.

## Dietary Factors

The dietary habits of overweight/obese adolescents in the MENA region, as documented by research studies, reveal incompatible results. For example, in Algeria, Allioua et al. conducted a study among adolescents aged 10–17 years and found that higher consumption of fatty food and the resulting unbalanced diet lead to higher BMI in both genders, as well as increased obesity and abdominal obesity among adolescent girls. This result is in line with another study (El-Gilany and El-Masry 2011) which reported that starchy food, fast food, snacks, and sweets thrice or more per week are among the main factors associated with increased risk. On the other hand, consumptions of vegetables, fruits, plant protein, and dairy products thrice or more per week reduced the risk. A study that examined Egyptian adolescents (12–15 years) showed that skipping breakfast, frequently eating snacks as a substitute, higher intake of fast food, as well as frequently snacking while watching television all lead to increased BMI. In Kuwait, adolescent males and females exhibited mixed dietary behavior, where consumptions of breakfast and fast food (in boys and girls), vegetables (in boys), potatoes, cakes and doughnuts, and sweets (in girls) were found significant predictors of overweight and obesity. Other studies in the MENA region found a significant positive correlation of the frequent consumption of unhealthy food, including fast food, sugar-sweetened beverages, carbonated/soft drinks, and sweets/candy (Amin et al. 2008; Raja'a and Mohanna 2005; Nasreddine et al. 2014; Pirinçci et al. 2010; Al-Hazzaa et al. 2012; El Kabbaoui et al. 2018), lack of dairy intake (Amin et al. 2008; Al Junaibi et al. 2013), infrequent consumption of breakfast and vegetables (Al-Hazzaa et al. 2012; Amin et al. 2008), inadequate energy intake (Jildeh et al. 2011), and energy consumption from carbohydrates (Nasreddine et al. 2010) and irregular snacking (Aounallah-Skhiri et al. 2008; Pirinçci et al. 2010) with excess weight.

Eating during school breaks and bringing food from home were protective factors against overweight/obesity among Bahraini female adolescents, whereas eating breakfast at home, eating in-between breakfast and lunch, as well as lunch and dinner reduced the risk among males (Musaiger et al. 2014). Also, consuming fruits thrice or more per week lessened the risk among males.

In contrast, the same study (Musaiger et al. 2014) reported no association of fast food intake and soft drink size (in both males and females) and consumption of vegetables, meat, seafood, legumes, dairy products, canned juices, and chocolates with overweight/obesity. Furthermore, the study by Nasreddine et al. (2014) found a negative association of obesity with higher dairy intake and regular breakfast consumption in Lebanon. In Qatar, Kerkadi et al. (2019) documented a negative association between obesity and the consumption of unhealthy food.

#### Lifestyle Factor

An association between obesity and reduced physical activity was documented by studies conducted in Algeria (Allioua et al. 2015), Egypt (El-Gilany and El-Masry 2011; Talat and El Shahat 2016), Iran (Abiri et al. 2019), Kuwait (Al-Haifi et al. 2013), Palestine (Jildeh et al. 2011), Saudi Arabia (Al-Hazzaa et al. 2012), and Tunisia (Zarrouk et al. 2009). However, Nasreddine et al. reported higher physical activity as moderately associated with lower odds of overweight on a sample adolescent population in Lebanon (Nasreddine et al. 2014). Also, the logistic regression analysis revealed no statistically significant relationship between overweight/obesity prevalence and physical activity (El Kabbaoui et al. 2018). Similar findings were detected by Kerkadi et al. (2019) and Al Junaibi et al. (2013).

Overweight/obesity prevalence was also found higher among adolescents who spent more time watching television (El-Gilany and El-Masry 2011; Talat and El Shahat 2016; Bener et al. 2011; Kerkadi et al. 2005) and using a computer (Abiri et al. 2019; El Kabbaoui et al. 2018), as compared to those who did not. In contrast, however, other studies have also identified no association between overweight/obesity and screen (Kerkadi et al. 2019; Al-Haifi et al. 2013).

Few studies reported that reduced sleep duration could contribute to weight gain in adolescents (Bener et al. 2011).

#### Other Risk Factors

El-Gilany et al. found that a history of obesity in one or both parents was an independent predictor of obesity among adolescents in Egypt (El-Gilany and El-Masry 2011). Similarly, the study by Nasreddine et al. (2010) and Raja'a and Mohanna (2005) documented that children of obese parents were at a higher risk of developing obesity. Furthermore, the onset of puberty among girls (Mikki et al. 2009) and duration of breastfeeding (Abiri et al. 2019) were also identified as significant obesity risk factors for adolescents in the region.

## 5.4 Discussion and Implications

The CDC designates overweight for children and adolescents as a body mass index (BMI) cutoff point at 85th percentile or above, whereas a cutoff point at 95th percentile or above defines obesity (Centers for Disease Control 2019; Maiti et al. 2013). Extreme obesity is defined at a BMI higher than or equivalent to the 99th percentile for age (ACOG committee opinion 2017). The International Obesity Task Force (IOTF) uses an international BMI cutoff point for specific age and sex to identify overweight and obesity for children and adolescents from age 2 to 18, while the cutoff thresholds for adults are BMI of 25 or above for "overweight" and 30 or above "obesity" (Cole et al. 2000; Maiti et al. 2013). These three most commonly adopted references have several drawbacks. First, these BMI benchmarks or references may not describe optimal growth given the extent of the positive skewness in body weight, which, as a result, may underestimate obesity among adolescents (Butte et al. 2007). Wang et al. (2006) also argue that the current reference data does not accurately reflect populations worldwide. Ethnic variations and the proportion of body fat correlated with adverse health effects have not been considered, as these references were developed on the basis of data collected in one or few developed countries.

There are various contributors to racial or ethnic differences in obesity metabolic comorbidities that might impact fat distribution, resting metabolic rate, insulin secretion and response, and lipids and lipoproteins (Akhabue et al. 2018). For example, adults and children in African American populations have lower visceral and hepatic fat as compared to White and Hispanic populations (Akhabue et al. 2018). Furthermore, White children have higher insulin sensitivity as compared with African American and Hispanic children, while higher rates of basal lipolysis are detected in White cohorts as compared to African American cohorts. These factors play an important role in the development and onset of obesity and its associated comorbidities and should be considered in quantitative assessment and predictive models (Akhabue et al. 2018). In addition, children and adolescent African Americans often have lower rates of adiponectin as compared to white Americans, which may help to better understand their elevated incidence of diabetes and cardiovascular disease despite the lower visceral adiposity (Akhabue et al. 2018).

While the literature indicates that there are several biological differences in the progression of obesity and the development of comorbidities across different races/ethnic groups, the correlations up to date remain far from conclusive. Researchers reviewed the growth data of healthy children in five major geographic regions of Africa, East Asia, South Asia, West Asia, and Europe in comparison with the NCHS/WHO reference. The studied children in most of these regions did not achieve heights similar to the NCHS/WHO reference medians (Butte et al. 2007).

On the other hand, according to Wang et al. (2019), even though obesity could be determined by genetic predisposition and ethnicity, most adolescent obesity results from the lack of physical activity and the consumption of more calories than needed for activity level. Several studies also remarked that the mean height of children differed slightly across ethnic groups relative to the socioeconomic variations within a specific ethnic community for prosperous populations. Furthermore, child development was largely determined by socioeconomic status and not by ethnicity or race in developed countries (Wang et al. 2019).

Although BMI remains the most common assessment tool for obesity, there are various other approaches available for assessing childhood and adolescent obesity, such as skinfold measurement of the visceral skinfold or that of the triceps muscles. Moreover, there are other methods that could be implemented in specific settings and are more accurate than BMI, including bioelectrical impedance, underwater weighing, dual-energy X-ray absorptiometry, and nuclear magnetic resonance (Centers for Disease Control and Prevention; Serra-Majem et al. 2007). Although these techniques can provide more accurate results of body fat and inform better on the likelihood of obesity-related health issues, they are costly, intrusive, not readily accessible, or challenging to standardize through observers or machines (Centers for Disease Control and Prevention).

The main cause for obesity at any age group, including adolescents, is positive energy balance, where the calories consumed are higher than those expended, resulting in fat accumulation and excess body weight (Narciso et al. 2019; Kadouh and Acosta 2017; Hruby and Hu 2015; Palou et al. 2000). As suggested by the (World Health Organization 2020b), a poor diet that is high in "empty" calories, including fast food and sugar-sweetened beverages, along with sedentary behaviors is a significant contributor to adolescent obesity. Multiple studies in fact concur that dietary energy density (Abdul-Rasoul 2012), and low levels of physical activity (Li et al. 2019; Han et al. 2010) are strongly associated with increased body fat in children and adolescents. However, attributing obesity and excess weight to a single or two factors, such as diet and physical activity for adolescent obesity, can be inaccurate and misleading (Ang et al. 2013).

## 5.5 Conclusive Remarks and Future Direction: Toward a Population-Specific Predictive Model

The concept of predictive modeling is well explored by researchers within a healthcare context. For example, Boukenze et al. developed a decision tree (C4.5)-based learning algorithm to predict chronic kidney disease (Boukenze et al. 2016), Kara et al. used artificial neural networks for diagnosing optic nerve disease (Kara et al. 2006), and Maclin et al. adopted a similar approach for cancer diagnosis (Maclin et al. 1991).

The high prevalence of overweight/obesity among adolescents in developed as well as developing countries, including the MENA region, and the resulting growing health and socioeconomic burdens urgently call for early intervention programs and effective prevention strategies. It is thus important to identify children at high risk of developing obesity later in life so that targeted intervention plans can be devised and implemented. Predictive algorithms have seen remarkable advancements in projecting adolescent obesity, utilizing both parental and childhood clinical data. Although obesity is associated with genetic and familial determinants, other elements, including social and environmental factors, contribute predominantly to the current pandemic as illustrated by various studies (Farrag et al. 2017).

Although predictive modeling has been partially implemented in the healthcare sector of some developed countries, specifically for predicting childhood/adolescent/adult overweight/obesity as illustrated in Table 1, there remains a significant scarcity of research that addresses the applicability of such models for the MENA region, where the higher prevalence and earlier onset of obesity are alarming.

The details of the available predictive algorithms for obesity prediction and the respective predictors are provided in Table 1. It can be emphasized that several models have reached good to excellent predictive performance, up to the level of clinical acceptance, when validated with different study cohorts (Pei et al. 2013; Graversen et al. 2015). These models have incorporated both modifiable and non-modifiable risk factors, including birth weight and childhood BMI (preschool BMI and current BMI) as strong predictors, along with other features including maternal data (BMI and smoking habits), sociodemographic factors, dietary, and environmental factors (Table 1). Although these factors yielded better performance, they cannot be directly adopted for the MENA region due to the potential impact of social, cultural, and religious factors. As such, Table 2 summarizes the particular risk factors as documented by the studies specific to the MENA.

Choosing the right set of input predictors or risk factors is critical as it helps to determine the predictive capability of a model. As highlighted in Table 2, sociodemographic factors (age, gender, family income, and parent's educational level and employment), physical activity, diet, screen time, parental obesity, family history of obesity, and duration of sleep are the most important risk factors applicable to the MENA population as revealed by various studies. By combining the findings from Table 1 and Table 2, we propose to consider physical activity, gender, parents' education/occupation, family income, parental obesity, diet, screen time, birth weight, current and childhood BMI, and maternal BMI as the potential predictors in our model.

Machine learning-based prediction techniques have emerged as effective tools to decipher and model big data, in recent years; however, scarcity of clinical data presents a major challenge to its successful implementation. It is thus imperative to initially propose logistic regression-based techniques for the MENA region, for which the performance has already been assessed and proved successful through validation on various cohorts (Rautiainen and Äyrämö 2019). Also, it is to be noted that, prior to any implementation, proper validation of the model considering a set of significant risk factors should be performed.

In conclusion, addressing adolescent obesity is critical toward minimizing its numerous associated chronic health risks. Since childhood and adolescent obesity are predictive of obesity in adults, special attention should be given to predictive models that could be invaluable for early intervention strategies aiming at combating this major public health burden. In this chapter, a predictive model specifically for adolescents in the MENA region had been proposed based on several main risk factors. Future work includes exploring machine learning-based powerful prediction algorithms for adolescent obesity and overweight by acquiring and incorporating a population-specific vast amount of clinical data. Unlike many other countries, adolescents in some countries in the MENA region, including the UAE, come from diverse ethnic backgrounds. It is, therefore, very important to investigate the possible contributions of social risk factor determinants to adolescent obesity. Future research also needs to examine the various predictive algorithms in the context of different obesity indicators, in addition to BMI (e.g., abdominal obesity and percent body fat), to provide more integrative comprehensive and valid predictive models for adolescent obesity.

## Reflection Questions

1. What are the multifactorial aspects and risk factors of obesity/overweight among adolescents within the MENA region?
2. Can previously developed population-based predictive models be suitable for estimating overweight/obesity among adolescents specifically within the MENA region?
3. What are the contributors to racial or ethnic differences in the progression of adolescent obesity and the development of comorbidities?
4. Can obesity predictive models address health and socioeconomic burdens in the future through early intervention programs and effective prevention strategies?
5. Can machine learning improve the applicability of predictive models for adolescent obesity through utilizing vast amount of clinical data and projecting analysis for a larger sample?

## References

Abdul-Rasoul, M. M. 2012. Obesity in children and adolescents in Gulf countries: facts and solutions. *Avances en diabetologia,* 28, 64–69.

Abiri, B., Sarbakhsh, P. & Vafa, M. 2019. Prevalence of overweight, obesity, and associated risk factors in healthy female adolescents in Tehran, Iran. *Central Asian Journal of Global Health,* 8.

ACOG Committee Opinion. 2017. *Obesity in Adolescents* [Online]. Available: https://www.acog.org/clinical/clinical-guidance/committee-opinion/articles/2017/09/obesity-in-adolescents.

Akhabue, E., Perak, A. M., Chan, C., Greenland, P. & Allen, N. B. 2018. Racial Differences in Rates of Change of Childhood Body Mass Index and Blood Pressure Percentiles. *The Journal of pediatrics,* 202, 98–105. e6.

Al-Haifi, A. R., Al-Fayez, M. A., Al-Athari, B. I., Al-Ajmi, F. A., Allafi, A. R., Al-Hazzaa, H. M. & Musaiger, A. O. 2013. Relative contribution of physical activity, sedentary behaviors, and dietary habits to the prevalence of obesity among Kuwaiti adolescents. *Food and nutrition bulletin,* 34, 6–13.

Al-Hazzaa, H. M., Abahussain, N. A., Al-Sobayel, H. I., Qahwaji, D. M. & Musaiger, A. O. 2012. Lifestyle factors associated with overweight and obesity among Saudi adolescents. *BMC public health,* 12, 354.

Al Hammadi, H. & Reilly, J. 2019. Prevalence of obesity among school-age children and adolescents in the Gulf cooperation council (GCC) states: a systematic review. *BMC obesity,* 6, 3.

Al Junaibi, A., Abdulle, A., Sabri, S., Hag-Ali, M. & Nagelkerke, N. 2013. The prevalence and potential determinants of obesity among school children and adolescents in Abu Dhabi, United Arab Emirates. *International Journal of Obesity,* 37, 68–74.

Alblooshi, A., Shaban, S., Altunaiji, M., Fares, N., Alshehhi, L., Alshehhi, H., Almazrouei, A. & Souid, A. K. 2016. Increasing obesity rates in school children in United Arab Emirates. *Obesity science & practice,* 2, 196–202.

Allioua, M., Djaziri, R., Mahdad, M. Y., Gaouar, S. B. S., Derradji, H., Boudjemaa, B. M. & Belbraouet, S. 2015. Dietary fat intake, micronutritient and obesity among adolescent in tlemcen (western Algeria). *Food and Nutrition Sciences,* 6, 860.

Amin, T. T., Al-Sultan, A. I. & Ali, A. 2008. Overweight and obesity and their association with dietary habits, and sociodemographic characteristics among male primary school children in

Al-Hassa, Kingdom of Saudi Arabia. *Indian journal of community medicine: official publication of Indian Association of Preventive & Social Medicine,* 33, 172.
Ang, Y. N., Wee, B. S., Poh, B. K. & Ismail, M. N. 2013. Multifactorial influences of childhood obesity. *Current Obesity Reports,* 2, 10–22.
Aounallah-Skhiri, H., Romdhane, H. B., Traissac, P., Eymard-Duvernay, S., Delpeuch, F., Achour, N. & Maire, B. 2008. Nutritional status of Tunisian adolescents: associated gender, environmental and socio-economic factors. *Public health nutrition,* 11, 1306–1317.
Baniissa, W., Radwan, H., Rossiter, R., Fakhry, R., Al-Yateem, N., Al-Shujairi, A., Hasan, S., Macridis, S., Farghaly, A. A. & Naing, L. 2020. Prevalence and determinants of overweight/obesity among school-aged adolescents in the United Arab Emirates: a cross-sectional study of private and public schools. *BMJ open,* 10, e038667.
Bener, A., Al-Mahdi, H. S., Ali, A. I., Al-Nufal, M., Vachhani, P. J. & Tewfik, I. 2011. Obesity and low vision as a result of excessive Internet use and television viewing. *International journal of food sciences and nutrition,* 62, 60–62.
Boukenze, B., Mousannif, H. & Haqiq, A. 2016. Predictive analytics in healthcare system using data mining techniques. *Comput Sci Inf Technol,* 1, 1–9.
Butler, É. M., Derraik, J. G., Taylor, R. W. & Cutfield, W. S. 2018. Prediction models for early childhood obesity: Applicability and existing issues. *Hormone research in paediatrics,* 90, 358–367.
Butte, N. F., Garza, C. & De Onis, M. 2007. Evaluation of the feasibility of international growth standards for school-aged children and adolescents. *The Journal of nutrition,* 137, 153–157.
Centers for Disease Control. 2019. *Childhood Obesity Facts* [Online]. Available: https://www.cdc.gov/obesity/data/childhood.html.
Centers for Disease Control. 2020. *Defining Adult Overweight and Obesity* [Online]. Available: https://www.cdc.gov/obesity/adult/defining.html [Accessed July 14, 2020.
Centers for Disease Control and Prevention. *Body mass index : considerations for practitioners* [Online]. Available: https://www.cdc.gov/obesity/downloads/bmiforpactitioners.pdf.
Cohen, I. G., Amarasingham, R., Shah, A., Xie, B. & Lo, B. 2014. The legal and ethical concerns that arise from using complex predictive analytics in health care. *Health affairs,* 33, 1139–1147.
Cole, T. J., Bellizzi, M. C., Flegal, K. M. & Dietz, W. H. 2000. Establishing a standard definition for child overweight and obesity worldwide: international survey. *Bmj,* 320, 1240.
Discigil, G., Tekin, N. & Soylemez, A. 2009. Obesity in Turkish children and adolescents: prevalence and non-nutritional correlates in an urban sample. *Child: care, health and development,* 35, 153–158.
El-Gilany, A.-H. & El-Masry, R. 2011. Overweight and obesity among adolescent school students in Mansoura, Egypt. *Childhood Obesity (Formerly Obesity and Weight Management),* 7, 215–222.
El Kabbaoui, M., Chda, A., Bousfiha, A., Aarab, L., Bencheikh, R. & Tazi, A. 2018. Prevalence of and risk factors for overweight and obesity among adolescents in Morocco. *Eastern Mediterranean Health Journal,* 24.
Farrag, N. S., Cheskin, L. J. & Farag, M. K. 2017. A systematic review of childhood obesity in the Middle East and North Africa (MENA) region: prevalence and risk factors meta-analysis. *Advances in pediatric research,* 4.
Graversen, L., Sørensen, T. I., Gerds, T. A., Petersen, L., Sovio, U., Kaakinen, M., Sandbaek, A., Laitinen, J., Taanila, A. & Pouta, A. 2015. Prediction of adolescent and adult adiposity outcomes from early life anthropometrics. *Obesity,* 23, 162–169.
Han, J. C., Lawlor, D. A. & Kimm, S. Y. 2010. Childhood obesity. *The lancet,* 375, 1737–1748.
Health Authority Abu Dhabi. 2016. *Health Statistics* [Online]. Available: www.hadd.ae.
Hruby, A. & Hu, F. B. 2015. The epidemiology of obesity: a big picture. *Pharmacoeconomics,* 33, 673–689.
Hudda, M. T., Fewtrell, M. S., Haroun, D., Lum, S., Williams, J. E., Wells, J. C., Riley, R. D., Owen, C. G., Cook, D. G. & Rudnicka, A. R. 2019. Development and validation of a prediction

model for fat mass in children and adolescents: meta-analysis using individual participant data. *bmj,* 366, l4293.

Ivanescu, A. E., Li, P., George, B., Brown, A. W., Keith, S. W., Raju, D. & Allison, D. B. 2016. The importance of prediction model validation and assessment in obesity and nutrition research. *International journal of obesity,* 40, 887–894.

Jator, E. K. 2014. Predicting Obesity among Adolescents in the United States Using Modified Logistic Model. *American Journal of Public Health,* 2, 86–90.

Jildeh, C., Papandreou, C., Mourad, T. A., Hatzis, C., Kafatos, A., Qasrawi, R., Philalithis, A. & Abdeen, Z. 2011. Assessing the nutritional status of Palestinian adolescents from East Jerusalem: a school-based study 2002–03. *Journal of tropical pediatrics,* 57, 51–58.

Kadouh, H. C. & Acosta, A. 2017. Current paradigms in the etiology of obesity. *Techniques in Gastrointestinal Endoscopy,* 19, 2–11.

Kara, S., Güven, A. & Öner, A. Ö. 2006. Utilization of artificial neural networks in the diagnosis of optic nerve diseases. *Computers in Biology and Medicine,* 36, 428–437.

Kerkadi, A., Abo-Elnaga, N. & Ibrahim, W. 2005. Prevalence of overweight and associated risk factors among primary female school children in Al-Ain city United Arab Emirates. *Emirates Journal of Food and Agriculture,* 43–56.

Kerkadi, A., Sadig, A. H., Bawadi, H., Al Thani, A. A. M., Al Chetachi, W., Akram, H., Al-Hazzaa, H. M. & Musaiger, A. O. 2019. The relationship between lifestyle factors and obesity indices among adolescents in Qatar. *International Journal of Environmental Research and Public Health,* 16, 4428.

Kim, C., Costello, F. J., Lee, K. C., Li, Y. & Li, C. 2019. Predicting Factors Affecting Adolescent Obesity Using General Bayesian Network and What-If Analysis. *International journal of environmental research and public health,* 16, 4684.

Li, N., Zhao, P., Diao, C., Qiao, Y., Katzmarzyk, P. T., Chaput, J.-P., Fogelholm, M., Kuriyan, R., Kurpad, A. & Lambert, E. V. 2019. Joint associations between weekday and weekend physical activity or sedentary time and childhood obesity. *International Journal of Obesity,* 43, 691–700.

Maclin, P. S., Dempsey, J., Brooks, J. & Rand, J. 1991. Using neural networks to diagnose cancer. *Journal of medical systems,* 15, 11–19.

Maiti, S., De, D., Ali, K. M., Bera, T. K., Ghosh, D. & Paul, S. 2013. Overweight and obesity among early adolescent school girls in urban area of West Bengal, India: prevalence assessment using different reference standards. *International journal of preventive medicine,* 4, 1070.

Mckinsey Global Report. 2015. *The Global Obesity Threat* [Online]. Available: www.Mckinsy.com.

Mikki, N., Abdul-Rahim, H. F., Awartani, F. & Holmboe-Ottesen, G. 2009. Prevalence and sociodemographic correlates of stunting, underweight, and overweight among Palestinian school adolescents (13-15 years) in two major governorates in the West Bank. *BMC public health,* 9, 485.

Morandi, A., Meyre, D., Lobbens, S., Kleinman, K., Kaakinen, M., Rifas-Shiman, S. L., Vatin, V., Gaget, S., Pouta, A. & Hartikainen, A.-L. 2012. Estimation of newborn risk for child or adolescent obesity: lessons from longitudinal birth cohorts. *PloS one,* 7, e49919.

Musaiger, A., Al-Mannai, M., Al-Lalla, O., Saghir, S., Halahleh, I.,Benhamed, M.,Kalam, F. & Ali, E. 2013. Obesity among adolescents in five Arab countries; relative to gender and age. *Nutricion hospitalaria,* 28, 1922–1925.

Musaiger, A. O., Al-Mannai, M., Tayyem, R., Al-Lalla, O., Ali, E. Y., Kalam, F., Benhamed, M. M., Saghir, S., Halahleh, I. & Djoudi, Z. 2012. Prevalence of overweight and obesity among adolescents in seven Arab countries: a cross-cultural study. *Journal of obesity,* 2012.

Musaiger, A. O., Al-Roomi, K. & Bader, Z. 2014. Social, dietary and lifestyle factors associated with obesity among Bahraini adolescents. *Appetite,* 73, 197–204.

Narciso, J., Silva, A. J., Rodrigues, V., Monteiro, M. J., Almeida, A., Saavedra, R. & Costa, A. M. 2019. Behavioral, contextual and biological factors associated with obesity during adolescence: A systematic review. *PloS one,* 14, e0214941.

Nasreddine, L., Mehio-Sibai, A., Mrayati, M., Adra, N. & Hwalla, N. 2010. Adolescent obesity in Syria: prevalence and associated factors. *Child: Care, Health and Development,* 36, 404–413.

Nasreddine, L., Naja, F., Akl, C., Chamieh, M. C., Karam, S., Sibai, A.-M. & Hwalla, N. 2014. Dietary, lifestyle and socio-economic correlates of overweight, obesity and central adiposity in Lebanese children and adolescents. *Nutrients,* 6, 1038–1062.

National Institute of Health. *Assessing Your Weight and Health Risk* [Online]. Available: https://www.nhlbi.nih.gov/health/educational/lose_wt/risk.htm [Accessed October 14, 2020.

Nouayti, H., Bouanani, N., Hammoudi, J., Mekhfi, H., Legssyer, A., Bnouham, M. & Ziyyat, A. 2020. Overweight and obesity in Eastern Morocco: Prevalence and associated risk factors among high school students. *Revue d'Épidémiologie et de Santé Publique.*

Palou, A., Serra, F., Bonet, M. & Pico, C. 2000. Obesity: molecular bases of a multifactorial problem. *European journal of nutrition,* 39, 127–144.

Pei, Z., Flexeder, C., Fuertes, E., Thiering, E., Koletzko, B., Cramer, C., Berdel, D., Lehmann, I., Bauer, C.-P. & Heinrich, J. 2013. Early life risk factors of being overweight at 10 years of age: results of the German birth cohorts GINIplus and LISAplus. *European journal of clinical nutrition,* 67, 855–862.

Pirinçci, E., Durmuş, B., Gündoğdu, C. & Açik, Y. 2010. Prevalence and risk factors of overweight and obesity among urban school children in Elazig city, Eastern Turkey, 2007. *Annals of human biology,* 37, 44–56.

Raja'a, Y. A. & Mohanna, M. A. B. 2005. Overweight and obesity among schoolchildren in Sana'a City, Yemen. *Annals of nutrition and metabolism,* 49, 342–345.

Rautiainen, I. & Äyrämö, S. 2019. Predicting overweight and obesity in later life from childhood data: A review of predictive modeling approaches. *arXiv preprint arXiv:1911.08361.*

Redsell, S. A., Weng, S., Swift, J. A., Nathan, D. & Glazebrook, C. 2016. Validation, optimal threshold determination, and clinical utility of the infant risk of overweight checklist for early prevention of child overweight. *Childhood Obesity,* 12, 202–209.

Riedel, C., Von Kries, R., Buyken, A. E., Diethelm, K., Keil, T., Grabenhenrich, L., Müller, M. J. & Plachta-Danielzik, S. 2014. Overweight in adolescence can be predicted at age 6 years: a CART analysis in German cohorts. *PloS one,* 9, e93581.

Serra-Majem, L., Ribas-Barba, L., Pérez-Rodrigo, C., Ngo, J. & Aranceta, J. 2007. Methodological limitations in measuring childhood and adolescent obesity and overweight in epidemiological studies: does overweight fare better than obesity? *Public Health Nutrition,* 10, 1112–1120.

Talat, M. A. & El Shahat, E. 2016. Prevalence of overweight and obesity among preparatory school adolescents in Urban Sharkia Governorate, Egypt. *Egyptian Pediatric Association Gazette,* 64, 20–25.

Waly, I., Zayed, K. & Al Haddabi, B. 2017. Obesity, eating habits and sedentary behaviour of Omani young adolescents: a cross-sectional study. *EC Nutrition,* 7, 3–10.

Wang, L., Perez, J., Heard-Costa, N., Chu, A. Y., Joehanes, R., Munson, P. J., Levy, D., Fox, C. S., Cupples, L. A. & Liu, C.-T. 2019. Integrating genetic, transcriptional, and biological information provides insights into obesity. *International journal of obesity,* 43, 457–467.

Wang, Y., Moreno, L. A., Caballero, B. & Cole, T. J. 2006. Limitations of the current world health organization growth references for children and adolescents. *Food and nutrition bulletin,* 27, S175–S188.

World Health Organization. 2018. *Adolescents: health risks and solutions* [Online]. Available: https://www.who.int/news-room/fact-sheets/detail/adolescents-health-risks-and-solutions.

World Health Organization. 2020a. *Adolescent health* [Online]. Available: https://www.who.int/health-topics/adolescent-health/#tab=tab_1 [Accessed July 14, 2020].

World Health Organization. April 2020b. *Obesity and overweight* [Online]. Available: https://www.who.int/news-room/fact-sheets/detail/obesity-and-overweight.

Zarrouk, F., Bouhlel, E., Feki, Y., Amri, M. & Shephard, R. J. 2009. Physical activity patterns and estimated daily energy expenditures in normal and overweight Tunisian schoolchildren. *Journal of sports science & medicine,* 8, 83.

# Chapter 6
# Determinants of Knowledge in Relation to Sexual and Reproductive Health of Adolescents in the Middle East and North Africa Region

**Doaa Oraby**

## 6.1 Introduction

The World Health Organization (WHO) defines an adolescent as an individual in the 10–19 years age group. Adolescence is a period of transition from childhood to adulthood and one of life's most complex stages, during which adolescents develop biologically and psychologically and move toward independence (WHO 2011). Adolescent sexual and reproductive health (ASRH) refers to the physical and emotional well-being of adolescents and includes their ability to remain free from unwanted pregnancy, unsafe abortion, sexually transmitted infections (STIs) including HIV/AIDS, and all forms of sexual violence and coercion. The 1994 International Conference on Population and Development established the basis for the advancement of ASRH (Plesons et al. 2019). Following this, ASRH has come to the forefront of the current global health agenda by its emphasis in two major global strategies: the 2030 Agenda for Sustainable Development (United Nations 2015a) and the United Nations Global Strategy for Women's, Children's, and Adolescents' Health (United Nations 2015b). Additionally, the Sustainable Development Goals (SDGs) provided the natural foundation for the vision of ASRH (United Nations 2016).

Despite the underscore of ASRH, much of the world refuses to acknowledge and accept adolescent sexuality, which is often discussed as a risk or problem to be avoided (Page and Shipley 2016). ASRH is still a low priority, and there are often restrictive laws and policies in addition to various societal, cultural, and religious factors that create an inhibitive environment for discussion of ASRH (Morris and

D. Oraby (✉)
Independent Reproductive Health Consultant, Cairo, Egypt



C. Barakat et al. (eds.), *Adolescent Health in the Middle East and North Africa*, Global Perspectives on Health Geography,
https://doi.org/10.1007/978-3-030-92107-1_6

Rushwan 2015). Adolescents become sexually active without accurate information on reproductive health, increasing the risk of having unintended pregnancies, unsafe abortions, and acquiring STIs including HIV (DeJong et al. 2007).

In the Middle East and North Africa (MENA) region, one in five people is an adolescent (WHO 2011). The region also hosts the largest number of refugees and displaced people in the world due to the prolonged emergencies in Libya, the Syrian Arab Republic, Yemen, and elsewhere, and more than one third of youth (15–24 years) live in fragile and conflict-affected countries (UNICEF 2019). In areas of conflict, many refugees and individuals live below the poverty line with limited opportunities that have negatively impacted their quality of life, and some of these resorted to health risk behaviors for survival (Al Buhairan 2015). Furthermore, young people caught in humanitarian crises face circumstances that can heighten their HIV vulnerability and other health threats due to generalized instability, violence, fractured communities, and collapsed health systems (ILO 2017). These demographic and socioeconomic transformations led to concerns about the rise of STIs, HIV, and unplanned pregnancies.

As adolescent SRH is becoming more prominent on the health agendas of many Arab countries of the MENA region, the current study aimed to identify the determinants of adolescents' knowledge of SRH. The literature search focused on following search terms "adolescents SRH in MENA OR adolescents SRH in Arab States OR young adults SRH in MENA OR young adults SRH in Arab States OR youth SRH in MENA OR youth SRH in Arab States," "sexual behavior of adolescents in MENA OR sexual behavior of adolescents in Arab States OR sexual behavior of young adults in MENA OR sexual behavior of young adults in Arab States OR sexual behavior of youth in MENA OR sexual behavior of youth in Arab States," and "Islam and adolescents SRH OR Islam and young adults SRH OR Islam and adolescents youth SRH." The author looked for background information related to the aforementioned search terms online by searching the Google Scholar and PubMed in addition to the websites of organizations involved in the delivery, funding, or evaluation of ASRH in resource-limited countries. These organizations included Family Health International, Guttmacher Institute, Interagency Youth Working Group, International Center for Research on Women, International Planned Parenthood Federation, Joint United Nations Programme on HIV and AIDS, Marie Stopes International, Pathfinder International, Population Council, United Nations Population Fund, United Nations Children's Fund, and World Health Organization (WHO). As to the search term of "Islam and adolescents SRH OR young adults SRH OR youth SRH," the author performed the online search in Arabic after failing to find adequate number of studies related to the topic in English. The results of the search displayed some Arabic textbooks that were available at the websites of social and religious universities and/or research institutions in MENA.

## 6.2 Results

### 6.2.1 *How Do Adolescents in MENA Learn About Sexual and Reproductive Health?*

***Parents:*** In MENA conservative culture, issues relating to SRH are considered taboos, and perceived stigma and embarrassment led to reluctance to discuss sexual health issues between parents and young people. The prevalent "culture of silence" related to SRH prevents parents and other adults from providing adolescents with the information that can help them with their emerging sexuality (Oraby and Hafez 2009). A nationally representative survey of young people and their parents in Egypt in the late 1990s reported that, although 42% of fathers reported talking to their adolescent sons about pubertal changes, only 7% of boys reported learning about puberty from their fathers (El-Tawila et al. 1999). Twenty years later, the situation did not differ; the 2010 Survey of Young People in Egypt (SYPE) – conducted with a nationally representative sample of about 15,000 young people between the age of 10 and 29 – demonstrated that parents do not discuss reproductive health and sexuality issues with their children, especially their sons. Family was reported as the source of information about puberty for 60.4% of females and only 5.8% of males. Among the male respondents, only 3.3% of the 10–14 years old, 6.5% of the 15–17 years old, and 9.1% of the 18–24 years old had talked with a parent about puberty. About a fourth (24.2%) of females aged 10–14, half (49.5%) of females aged 15–17, and 50.8% of those aged 18–24 had talked with a parent about puberty (Population Council 2010).

***School:*** In the MENA region, SRH education is generally perceived as a manifestation of the pervasive influence of the Western social norms that clash with Muslim or cultural values and norms about sex and gender, which could lead to premature sexualization and corruption through awakening sexual awareness (Farrag and Hayter 2014). In Oman, a study among a nationally representative sample of secondary school students of both sexes revealed that only half of the sample knew the changes at puberty of their own sex, while even fewer knew the changes in the opposite sex and knowledge of fertility period was poor (Jaffer et al. 2006). In most Egyptian classrooms, if SRH topics are covered, they are done so in biology and not covered in any detail. Other times, SRH sections are skipped or covered inadequately because teachers are unprepared or embarrassed (Geel 2012). School was cited as the source of information about puberty for 11.5% of male and 4.1% of female SYPE respondents (Population Council 2010).

***Internet:*** One in three Internet users around the world are children and adolescents aged younger than 18 years who independently seek SRH information on their own terms (UNICEF 2017). Unfortunately, adolescents who count on information from the Internet to fulfill their unmet need for SRH knowledge are often provided with inaccurate information. One study referred to an Arabic website mentioning that

HIV could be transmitted through saliva or a toothbrush (Benikrane 2013). Concurrently, the Internet comes with easy accessibility to online pornography with its impact on adolescent SRH. Exposure to online pornography is further aggravated by youth-limited capacity for self-control and susceptibility to peer pressure with subsequent potential compulsive patterns of online use (Griffiths 2001). A study among Egyptian youth, aged from 13 to 24, revealed that 74% of the participants have watched porn clips before, with a significant gender-based difference (males 92% vs. females 48.4%) (Menshawy et al. 2020). Another study conducted among pre-university young Saudi men (15–20 years) revealed that 61% reported watching pornographic movies and most of them (71%) viewed these movies through the Internet. Premarital sexual activity was associated with viewing of pornographic movies with 6.79 odds ratio (i.e., six times more than those who do not watch pornographic materials) (Raheel et al. 2013). Furthermore, studies revealed that young people who use sexual material online are more likely to engage in at-risk sexual behavior for HIV/AIDS (Benavides et al. 2012).

### 6.2.2 Patterns and Consequences of Sexual Behavior of Adolescents in MENA

***Premarital Sex:*** Several studies around the MENA region highlighted that quite a large proportion of young men are engaged in premarital sex, which goes against the assumption that youth might be reluctant to engage in such behaviors in a region with strong religious and moral objection to premarital and extramarital sex (Raheel et al. 2013). However, what is promoted religiously is not necessarily what is practiced, and counting only on the protective effect of religious and traditional values is not enough (Ridanovic 1997). Yet, the prevalence of premarital sexual relations is difficult to determine as many reports and studies often conflict one another, and the accuracy of related surveys may be difficult to determine due to the taboo nature of the subject. Furthermore, statistics for indicators such as hymen reconstruction surgery and backstreet abortions reflecting the existence of premarital sexual behavior are largely based on observations and anecdotal evidence (Bahgat and Afifi 2004).

A study conducted among youth (15–24 years) in Jordan showed that 7% of college students and 4% of youth among the general population reported to non-marital sex (Johns Hopkins University 2001). Another study among Lebanese university students revealed that the majority of males (73.3%) and one fifth of females (21.8%) declared previous sexual relations (Barbour and Salameh 2009). In Egypt, a survey of 10,000 respondents aged 13–35 revealed that 11.3% of young men and women knew “other” young people who were in a relationship with the opposite sex. Out of those who answered yes, 18.9% also indicated that these relationships were intimate. Authors noted that those figures may be an underreporting as interviews took place in respondents’ homes; thus, some young people may have been afraid of being overheard by family members (Sieverding and Ragab 2015).

Although marriage substitutes have always existed, their expansion has aligned with the changing terrain of sexual behavior (Nagi 2017). Urfi marriage is an unregistered form of marriage that some young people resort to aiming to give sexual relationships religious legitimacy. It consists of two people writing down on a piece of paper that they are married and is signed by two witnesses, usually friends of the couple, but it is otherwise kept a secret from their families and peers (Wynn 2016). In Egypt, a study involving 4566 young males and females aged 18–30 years showed that the prevalence of Urfi marriage was 4% among the total population of youth and 6% among university students, despite social stigmatization of the practice (El-Tawila and Khadr 2004).

***HIV/AIDS:*** The MENA region has the lowest HIV prevalence in the world (less than 0.1%). Limited HIV transmissions in the MENA region relative to other regions had been attributed to several protective factors, namely, male circumcision and Islamic cultural traditions. Male circumcision, which is associated with a 60% efficacy against HIV infection for men, is universally practiced in MENA (Bailey et al. 2007), and Islamic cultural traditions have been cited as a protective factor even after adjustment for male circumcision (Hargrove 2008). However, what is promoted religiously is not necessarily what is practiced, and counting only on the protective effect of religious and traditional values is not enough to prevent the progression of the HIV epidemic (Ridanovic 1997).

Since 2010, the annual number of new HIV infections (all ages) has declined by 16%, and AIDS-related mortality (all ages) has declined by 33% globally. However, the epidemic in MENA continues to grow, with a 10% increase in new infections and a 9% increase in the annual number of AIDS-related deaths between 2010 and 2018 (UNAIDS data 2019). The HIV vulnerability of young people in MENA has increased due to the economic decline, war, and civil unrest in the region in recent years. The vulnerability is further heightened by the low levels of knowledge of HIV modes of transmission and the preventive role of condom (Abdel-Tawab et al. 2016). A study conducted among pre-university young Saudi men (15–20 years) revealed that only 51% knew that condom use could prevent STIs, and 20% were not aware that HIV could be transmitted through both homosexual and heterosexual contacts (Raheel et al. 2013). The 2015 Egypt Health Issues Survey revealed that knowledge of AIDS is lowest among women and men aged 15–19 (50% and 57%), respectively, and that comprehensive knowledge about HIV[1] is extremely low among young women and men aged 15–24 (4.1% and 6.6%), respectively (MOHP et al. 2015).

---

[1] Comprehensive knowledge of HIV means knowing that consistent use of condoms during sexual intercourse and having just one uninfected faithful partner can reduce the chance of getting the AIDS virus, knowing that a healthy-looking person can have the AIDS virus, and rejecting two common local misconceptions about transmission or prevention of the AIDS virus.

### 6.2.3 How Can Islam Tackle the Shortcomings of Adolescents' SRH in MENA?

Islam, the prevalent religion in the region, is explicit about many aspects of human sexuality, and numerous hadith[2] show the Prophet's willingness to discuss these matters openly. Quran[3] has placed emphasis on acquiring knowledge, and in the days of Prophet Muhammad, Muslim men and women were never too shy to ask him questions related to sexual life. Islam encourages education about matters related to sex in a way that informs young people about sexuality in scientific and moral terms (Bouhdiba 2008; Khan et al. 2020). However, the culture of silence in the MENA countries considers sex a taboo to be addressed only before marriage. This could explain the controversial nature of educating young people about SRH being not a result of Islamic doctrine but rather the result of cultural sexual taboos misinterpreted as grounded in Islam (Tabatabaie 2015).

Islam prohibits premarital sex and what leads to it (anything which breaks down sexual inhibition and loss of self-control, i.e., alcohol, drugs, watching pornography). Islam promotes behavior conforming to several themes of STIs and HIV prevention, including prohibitions against premarital and extramarital sex (Lenton 1997; Bouhdiba 2008) and prohibitions against alcohol consumption, which is strongly associated with higher-risk behavior (Kaljee et al. 2005).

## 6.3 Conclusion and Recommendations

In the MENA region, adolescents constitute almost one fifth of the population, and more than one third of youth live in fragile and conflict-affected countries, which led to concerns about ASRH. There is lack of SRH education at schools, and parents/adults often have difficulty talking with adolescents about SRH owing to embarrassment or ignorance. Adolescents become sexually active without accurate information on SRH, which is a serious concern especially in light of the changing sexual terrain among young people. Adolescents' limited SRH information combined with easy accessibility of online pornography and their curiosity and immaturity heighten their odds of vulnerability to risky sexual behaviors and subsequent sexual ailments, including HIV/AIDS.

Sexual and reproductive health education programs are critical to provide adolescents accessibility to reliable sources of information. However, providing adolescents with SRH education is likely to be a difficult process due to the perceived encroachment of Western values. For example, the topics of premarital sex and homosexuality are pillars of ASRH programs implemented in Western countries but

[2] Record of the traditions or sayings of the Prophet Muhammad, revered and received as a major source of religious law and moral guidance.

[3] Holy book of Islam.

are frowned upon by MENA conservative communities, not only Muslims, and both topics are considered sinful in Islam. Hence, the Western approach to ASRH education should be tailored and culturally adapted to the MENA region because of the differences in history, religion, and culture. We can build on the experience of introducing HIV harm reduction interventions in Egypt in 2008, where the package of interventions was not implemented in full. The provision of opioid substitution therapy was not approved, yet the concept of harm reduction was not rejected but rather tailored to encompass (provision of HIV testing, condoms, and syringes). Harm reduction is supported by the two key tenets of Islam, namely, doing no harm to oneself or others, and the worst harm is eliminated by a lesser harm (Oraby 2013).

Islam recognized the power of sexual need and how it should be addressed, yet parents/adults are not familiarized with that. Hence, religious institutions should encourage pursuing valid scientific knowledge during Friday congregational prayers and via different media platforms. Concurrently, schools have potential to provide adolescents with credible information and have advantages in terms of access, equity, and responsiveness to adolescents' needs. School nurses, supervisors, and teachers' education should be tackled with special focus on communicating with adolescents on the core issues of risk-taking and sexual development.

Adolescent SRH education programs should also benefit from the popularity of social media. Facebook and Twitter can play an important role in the outreach of adolescents with reliable information and preventive messages and provide them with the opportunity of asking anonymous questions that they would be embarrassed to ask face to face. Mobile phone technology can also provide an opportunity to reach adolescents with SRH messages and lists of nearby places where they can access friendly ASRH services. Simultaneously, adolescents should be informed and empowered for wise use of social media, and parents should be informed, not shy away, from the topic of adolescents watching online pornography and alternatively acknowledge and resort to professional support. Accordingly, professional cadre of counselors who are competent to tackle the possible influences of pornography on sexual behavior, attitudes about sex, and relationships is needed.

Country-specific research is encouraged to study (1) adolescents' sexual knowledge, attitudes, and behaviors and their association with socioeconomic standards and other behaviors such as watching pornography, delinquencies, and drug use, (2) individual and contextual factors associated with watching online pornography, (3) the acceptance and uptake of social media for adolescents' SRH education, (4) extent and association of premarital sexual relations with rising levels of STIs and HIV among youth, and (5) exploring the reasons why parents/adults are not familiarized that Islam recognized the power of sexual need and how it should be addressed and testing the feasibility of potential interventions to address this shortcoming.

## Reflection Questions

- Is adolescent reproductive and sexual health still a low priority in MENA despite the heightened vulnerability of adolescents caused by economic decline, civil war, lack of accurate knowledge, and easy accessibility to online pornography?
- How will the HIV epidemic in MENA region progress given the economic decline, war, and civil unrest? (The region hosts the largest number of refugees and displaced people, many of whom live below the poverty line and resort to risky behaviors.)
- Do we have to adopt the Western approach to adolescent sexual and reproductive health as is? (The HIV harm reduction interventions were tailored and culturally adapted to the MENA region.)
- How does Islam perceive education of adolescents about matters related to sex?
- How can we tackle the controversial nature of educating young people about SRH being not a result of Islamic doctrine but rather the result of cultural sexual taboos?

## References

Abdel-Tawab, N., Oraby D., Saher S., and Ismail S. 2016. "Understanding HIV-Related Vulnerabilities and Stigma Among Egyptian Youth," Cairo: Population Council – Egypt Office.

Al Buhairan F. Adolescent and Young Adult Health in the Arab Region: Where We Are and What We Must Do. Editorial / Journal of Adolescent Health 57 (2015) 249–251

Bahgat, H. and Afifi, W. 2004. Sexuality politics in Egypt. In R. Parker, R. Petchesky, & R. Sember (Eds.), Sex politics: Reports from the front lines (pp. 53–90). Washington, DC: Sexuality Policy Watch.

Bailey R C, Moses S, Parker CB, Agot K, Maclean I, Krieger JN, et al. Male Circumcision for HIV Prevention in Young Men in Kisumu, Kenya: A Randomised Controlled Trial. Lancet 2007, 369: 643–56.

Barbour B, Salameh P. Knowledge and practice of university students in Lebanon regarding contraception. Eastern Mediterr Health J 2009;15(2):387–99.

Benavides R., Montero V., González V., and Rodríguez O., 2012. Use of Sexual Material Online and At-Risk Sexual Behavior Regarding HIV/AIDS among College Students. Nurs Inform. 2012; 2012: 021.

Benikrane, M. 2013. "Online information on youth sexual and reproductive health available in Arabic: A content analysis from Egypt." Cairo: Ford Foundation (Unpublished report.)

Bouhdiba A. 2008. Sexuality in Islam, New York, Routledge, 2008.

DeJong J, Shepard B., Roudi-Fahimi F., and Ashford L. 2007. "Young people's sexual and reproductive health in the Middle East and North Africa," Reproductive Health 14(78): 8.

El-Tawila S., Ibrahim B., Sallam S., El Gibaly O. and El Sahn F., 1999. Transitions to Adulthood: A National Survey of Egyptian Adolescents. Cairo: Population Council.

El-Tawila, S. and Khadr, Z. 2004. "Patterns of Marriage and Family Formation among youth on Egypt." National Population Council, Research Management Unit and Center for Information and Computer Systems, Faculty of Economics and Political Science, Cairo University, Cairo Egypt.

Farrag, S., and Hayter, M. 2014. A qualitative study of Egyptian school nurses' attitudes and experiences toward sex and relationship education. The Journal of School Nursing, 30 (1), 49–56.

Geel F. 2012. Quality sexual education needed for adolescents in Egyptian schools. Washington, DC: Population Reference Bureau.

Griffiths M. 2001 'Sex on the internet: Observations and implications for internet sex addiction', Journal of Sex Research, 38: 4, 333–342.

Hargrove J. Migration, mines and mores: the HIV epidemic in southern Africa. South Afr J Sci 2008, 104, Issue 1 & 2:53–61.

International Labor Office (ILO) 2017. Promoting a rights-based approach to migration, health and HIV and AIDS: a framework for action. Geneva: International Labor Office.

Jaffer Y.A., Afifi M., Al Ajmi F. and Alouhaishi K. Knowledge, attitudes and practices of secondary-school pupils in Oman: II. Reproductive health. Eastern Mediterranean Health Journal, Vol. 12, Nos 1/2, 2006

Johns Hopkins University, Center for Communication Programs, National Population Commission, Jordan. Jordanian youth survey; knowledge, attitude and practices on reproductive health and life planning. Amman: Johns Hopkins University, 2001.

Kaljee LM, Genberg B L, Minh TT, Tho LH, Thoa LT, Stanton B. Alcohol Use and HIV Risk Behaviors among Rural Adolescents in Khanh Hoa Province VietNam. Health Educ Res 2005, 20: 71–80.

Khan M, Rassool G and Mabud S. 2020. Sexuality Education from an Islamic Perspective. Cambridge Scholars Publishing.

Lenton C. 1997. Will Egypt escape the AIDS epidemic? THE LANCET Vol 349. April 5, 1997.

Menshawy A, Abushouk AI, Ghanem E, Senna M, Lotfy N, Abdel-Maboud M, Elmaraezy A, Elsherbeny MY. Break the silence: knowledge and attitude towards sexual and reproductive health among Egyptian youth. Community Ment. Health J. 2020.

Ministry of Health and Population (MOHP)-Egypt, El-Zanaty and Associates [Egypt], and ICF International. Egypt Health Issues Survey 2015. Cairo and Rockville, MD: Ministry of Health and Population and ICF International.

Morris J.L. and Rushwan H. Adolescent sexual and reproductive health: The global challenges. International Journal of Gynecology and Obstetrics 131 (2015) S40–S42.

Nagi M. 2017. "Islam, Sexualities and Education." In International Handbook of Islamic Education, edited by HolgerDaunand Reza Arjmand, 1–12 New York: Springer.

Oraby D. and Hafez Z. 2009. Formative Assessment of Youth Reproductive Health Needs in Menofia and Ismailia Governorates. Family Health International and The United Nations Population Fund. Available at: http://www.fhi360.org/NR/rdonlyres/e7ke-fkqbh5rv4gnndbudjyplqepbnmn5grsxtdw6gjf7quptksvhiuehizzsdor3f6rzdvlkaqcvmo/FormativeAssessmentYouthRHNeedsRH.pdf

Oraby D. 2013. Harm reduction approach in Egypt: the insight of injecting drug users. Harm Reduct J 10, 17 (2013). https://doi.org/10.1186/1477-7517-10-17

Page SJ. and Shipley H (2016). Sexuality. In: Yamane D. (eds) Handbook of Religion and Society. Handbooks of Sociology and Social Research. Springer, Cham. https://doi.org/10.1007/978-3-319-31395-5_20

Plesons M. et al. 2019. Forward, Together: A Collaborative Path to Comprehensive Adolescent Sexual and Reproductive Health and Rights in Our Time. Journal of Adolescent Health 65 (2019) S51–S62.

Population Council. 2010. The Survey of Young People in Egypt. Cairo: Population Council.

Raheel H, Mahmood MA, BinSaeed A. Sexual practices of young educated men: implications for further research and health education in Kingdom of Saudi Arabia (KSA). J Public Health (Oxf) 2013; 35:21–6.

Ridanovic Z. AIDS and Islam. Medicinski Archiv 1997, 51: 45–46 1997, 51, 45–46.

Sieverding M. and Ragab A., 2015. "Marriage and Family Formation Trends among Youth in Egypt" In: Rania Roushdy and Maia Sieverding (eds) Panel Survey of Young People in Egypt (SYPE) 2014. Generating Evidence for policy, programs and research. Cairo: Population Council.

Tabatabaie A. (2015). Childhood and adolescent sexuality, Islam, and problematics of sex education: A call for re-examination. Sex Education, 15(3), 276–288.
UNAIDS Data 2019: accessed at http://aidsinfo.unaids.org/ on May 10, 2020.
United Nations 2015a. Transforming our world: The 2030 agenda for Sustainable development; 2015. Available at: https://sustainabledevelopment.un.org/content/documents/21252030%20Agenda%20for%20Sustainable%20Development%20web.pdf.
United Nations 2015b. Every woman, Every Child. Global Strategy for Women's, Children's, and adolescents' health (2016–2030): Survive, thrive, Transform; 2015. Available at: http://www.who.int/pmnch/media/events/2015/gs_2016_30.pdf.
United Nations 2016. About the sustainable development goals. New York: United Nations; 2016. Available at: https://www.un.org/sustainabledevelopment/sustainable-development-goals/.
United Nations Children's Fund (UNICEF). The state of the world's children 2017: Children in a digital world. New York: UNICEF; 2017. Available at: https://www.unicef.org/publications/files/SOWC_2017_ENG_WEB.pdf.
UNICEF 2019. MENA Generation 2030.Investing in children and youth today to secure a prosperous region tomorrow. Amman: UNICEF Regional Office for the Middle East and North Africa. Available at: https://www.unicef.org/mena/reports/mena-generation-2030
World Health Organization (WHO). Sixty-fourth World Health Assembly. Resolution WHA 64.28: Youth and health risks. Geneva: World Health Organization, 2011. http://apps.who.int/gb/ebwha/pdf_files/WHA64/A64_R28-en.pdf
Wynn L.L. 2016: 'Like a Virgin': Hymenoplasty and Secret Marriage in Egypt, Medical Anthropology, https://doi.org/10.1080/01459740.2016.1143822

# Part III
# Epidemiology of Adolescent Health in the MENA

# Chapter 7
# Growth Assessment in Adolescence: Measurement and Interpretation for the Middle East and North Africa

**Leila Cheikh Ismail and Ayesha Salem Al Dhaheri**

## Key Highlights

- Context
  - Adolescence is a critical phase characterized by rapid physical growth alongside various psychosocial, emotional, and cognitive aspects.
  - Proper assessment and monitoring of adolescence growth are keys toward healthy growth patterns and early identification of any abnormalities.
- Objectives
  - To provide an overview on the 2007 WHO growth charts as an assessment tool.
  - To discuss interpretations of growth parameters and their implications within the MENA region.
- Methodology
  - Basic growth parameters and proper standardized measurement techniques.

L. Cheikh Ismail (✉)
Department of Clinical Nutrition and Dietetics, College of Health Sciences, University of Sharjah, Sharjah, United Arab Emirates
e-mail: lcheikhismail@sharjah.ac.ae

A. S. Al Dhaheri
Department of Nutrition and Health, College of Medicine and Health Sciences, United Arab Emirates University, Al Ain, United Arab Emirates
e-mail: Ayesha_aldhaheri@uaeu.ac.ae

 
C. Barakat et al. (eds.), *Adolescent Health in the Middle East and North Africa*, Global Perspectives on Health Geography,
https://doi.org/10.1007/978-3-030-92107-1_7

  - Interpretation and proper use of growth charts as an assessment tool.
- Key Findings
  - The 2007 WHO growth references are the most reliable representative tools available to assess adolescent growth.
  - Many countries in the MENA region adopted the WHO Child Growth Standards. However, the use of the 2007 WHO adolescent's growth charts was not assessed.
- Conclusion and Implications
  - Limited data is available on adolescence growth assessment in the MENA region.
  - Standardized growth assessment methods and tools should be adopted in the MENA region to allow proper evaluation and comparison.

## 7.1 Introduction

Adolescence is one of the most challenging phases of human life, representing a transition from childhood to adulthood. The World Health Organization (WHO) defines adolescence between ages 10 and 19. They comprise about one sixth of the global population totaling 1.2 billion in 2019. While growth occurs at a gradual pace in early childhood, a more rapid pattern of growth characterizes adolescence and is accompanied by changes in various developmental aspects, including physical, psychosocial, cognitive, and emotional growth.

Adolescence is thought to be a healthy phase of life by means of low prevalence of infectious diseases compared to infants and less chronic diseases compared to adults, resulting in less attention put toward this age group. Contrary to the large body of studies on childhood and adulthood, little data is available on adolescent growth and anthropometry. However, it requires more consideration as it manifests higher risks for unhealthy harmful behavior and irrational decisions associated with substance abuse, pregnancy, sexually transmitted disease, and injuries whether intentional or accidental (WHO 1995). Hence, due to the uniqueness of this phase, it is essential to assess and monitor nutritional status as it lays out the foundation of healthy growth and lifestyle behaviors toward adulthood.

Growth velocity in adolescence exceeds that of early childhood especially during the *growth spurt* that characterizes this phase, demanding higher nutrient requirements. It is the period where the highest percentage of height is achieved and starts between 8 and 13 years of age on average and reaches a maximum growth rate referred to as *peak height velocity* two years after its onset. Adolescents averagely gain 20% of their adult height accompanied by an increase of about 40–50% of their adult weight during puberty (Mahan and Raymond 2016). Puberty starts with a phase of rapid growth, followed by a phase of deceleration and eventually reaching cessation of growth by adulthood (WHO 1995). Patterns of growth such as maturation timing or tempo vary significantly among different settings. Differences are attributed to different factors such as discrepancy of maturation timing, sexual and skeletal maturation, genetics, nutritional status, environment, and socioeconomic level (Azzopardi et al. 2017).

### 7.1.1 *Measurements of Growth*

As adolescence is a period of hormone-mediated change and maturation, anthropometry provides the best means of monitoring physical growth and evaluating deviations, such as underweight, overweight, stunting, and wasting. Moreover, it provides an easy, noninvasive, and cost-effective means of assessment, which in turn contributes to early intervention strategies. Linear growth can be assessed using basic anthropometric measurements, including weight, height, and a derived value of body mass index (BMI).

In order to be able to interpret measurements, certain indices must be defined, for example, a weight measurement has to be compared to another, such as age or height, in order to have a meaning. Therefore, a derived value of weight over height squared also known as BMI is appropriate for data interpretation. Contrary to adults, no simple cutoff values exist for adolescents, thus age-adjusted parameters are used; including weight-for age, height-for-age and BMI for age (De Onis et al. 2007). These indices can be expressed as z-scores or percentiles to facilitate interpretation and comparison and serve as sensitive indicators to health, growth, and development. In line with the WHO recommendations, z-score will be adopted for interpretation and discussion (WHO 1995).

Anthropometric data should be discussed separately for each sex in the case of adolescence, considering the differences in maturation and growth spurt timing. Moreover, maturational status based on age should be taken into consideration when interpreting anthropometry reference data in order to obtain a clearer indication of physical growth and nutritional status (WHO 1995).

Assessment of nutritional status necessitates the use of appropriate measurement techniques and reliable tools. In 2007, the WHO merged data from the 1977 National Center for Health Statistics (NCHS)/WHO growth reference (1–24 years) with data from the under-fives growth standards' cross-sectional sample (18–71 months) to smooth the transition between the two samples. It developed the 2007 WHO growth references for school-aged children and adolescents (De Onis et al. 2007). The growth references allow measurements to be plotted accurately and periodically. They also provide insights into monitoring growth and nutritional health for adolescents, as well as provide accurate estimates of malnutrition in this critical developmental phase.

### *7.1.2 Standardization of Measurements*

Ensuring highest quality of measurement and minimizing data error requires adequate training, continuous standardization, and adherence to measurement procedures in data collection, which in turn allows proper interpretation of growth and nutritional status. It is critical to follow measurement techniques and guidelines from reliable anthropometry manuals, protocols, and literature (De Onis et al. 2004). The WHO applied standardization protocols in the process of producing growth references, primarily by performing standardization training sessions. Frequent standardization sessions are crucial to assess adherence to protocol, monitor accuracy and precision and quality control, and take corrective actions in case of deviation (WHO 2008). Moreover, instruments need to be accurate, precise, and calibrated on a regular basis (Cheikh Ismail et al. 2013).

Youth population aging 10–24 years makes up approximately a third of the total population according to the World Bank, Global Burden of Disease Study in 2015 (Sawyer et al. 2018). The fact that adolescents encompass a relatively large proportion of society dictates the urgent need to focus on this group and address their health needs. Unfortunately, inadequate research addressed the health risk and problems of this age group within the MENA region. This forms a significant barrier to formulate a comprehensive understanding of their overall health and nutritional status and impede the development of appropriate policies and strategies to address and improve their physical and mental health.

In this chapter, an overview of the 2007 WHO growth charts as an assessment tool will be provided with a discussion of the interpretation of growth patterns in health and disease among adolescents and implications of adolescent growth assessment using the new WHO references within the MENA region.

## 7.2 Methodology and Interpretation

### *7.2.1 Measurements*

#### Measuring Weight

Weight is an essential measurement and variable used in energy calculations and body composition indices. A platform electronic scale calibrated to 0.1 kg is appropriate to use for adolescents. Minimal clothing is advised, and any footwear should be removed. The subject should stand on the center of the scale with weight distributed equally on both feet. Weight is recorded to the nearest 0.1 kg. Multiple readings are required for verification and to avoid error. Measurements are then plotted on the growth charts for interpretation and comparison.

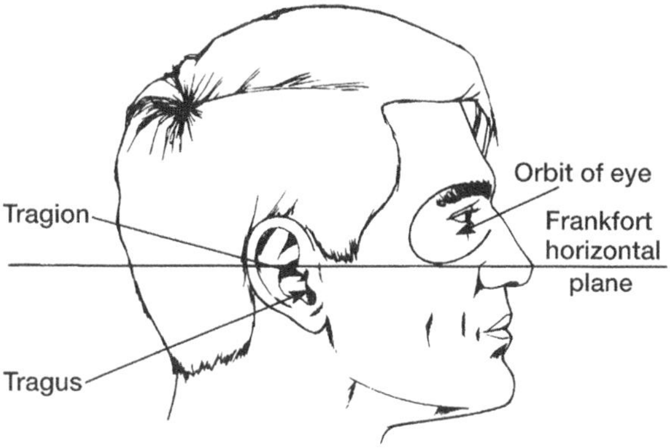


**Fig. 7.1** Horizontal Frankfurt plane: the line between the auditory canal and the corner of the eye should be horizontal to the floor, i.e., in a Frankfurt horizontal plane. (Source: Gibson 2005)

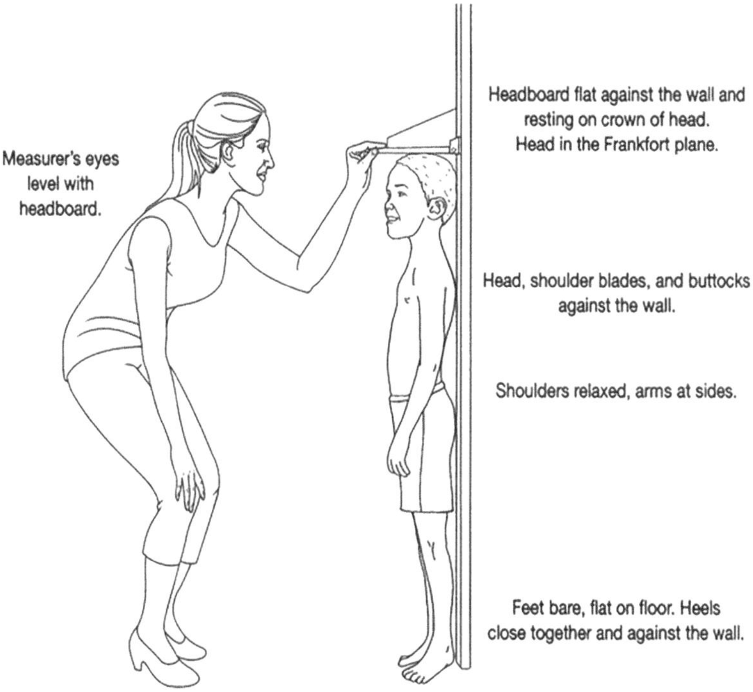


**Fig. 7.2** Standardized position of body segments to measure height. (Source: Gibson 2005)

## Measuring Height

Height is best measured using a Harpenden stadiometer. The subject should be barefoot, and any headwear or hair ornament should be removed for appropriate positioning. The heels should be slightly apart and flat on the platform, arms on the sides, legs straight, shoulders relaxed, and head in Frankfort horizontal plane as depicted in Fig. 7.1 (Gibson 2005).

The body should be in contact with the vertical backboard of the stadiometer in multiple points as shown in Fig. 7.2. To ensure best measurement, the subject should take a deep breath and stand erectly before reading. Measurements are recorded to the nearest 0.1 cm.

#### Body Mass Index

Body mass index (BMI) refers to the ratio of the body mass divided by squared height and is derived using the following equation:

$$BMI = \frac{\text{weight}(\text{kg})}{(\text{height}(\text{m}) * \text{height}(\text{m}))}$$

The weight must be measured in kilograms (kg) and the height in meters (m).

**Example A person weighing 45 kg and who is 150 cm tall has the BMI of 20 kg/m$^2$. Firstly, the height should be converted to meters (1.5 m), and then follow the equation to obtain the BMI value.**

### *7.2.2 Assessment Tools*

Growth curves are a product of data from a large number of adolescents from various ages. Previously, the 1977 National Center for Health Statistics (NCHS)/WHO international growth reference was recommended by the WHO for children above 5 years; however, it had some shortcomings. In order to fill the gap in growth charts and provide references for the 5–19 years age group, the WHO developed new growth references for adolescents aligned with the WHO Child Growth Standards at 5 years and the recommended adult cutoffs for overweight and obesity at 19 years (De Onis et al. 2007). Figure 7.3 shows an example of the 2007 WHO growth charts. Separate reference charts and tables are available for boys and girls. The WHO growth reference (5–19 years) is as follows:

- Height for age (5–19 years)
- Weight for age (5–10 years)
- BMI for age (5–19 years)

It should be noted that the height-for-age and BMI-for-age charts are available for the ages 5–19 years, whereas weight-for-age charts extend to only 10 years due to the limitation of not being able to differentiate between height and body mass, where height is sensitive and rapidly changing during puberty growth spurt (i.e., a reading may be interpreted as being overweight although the individual may be just

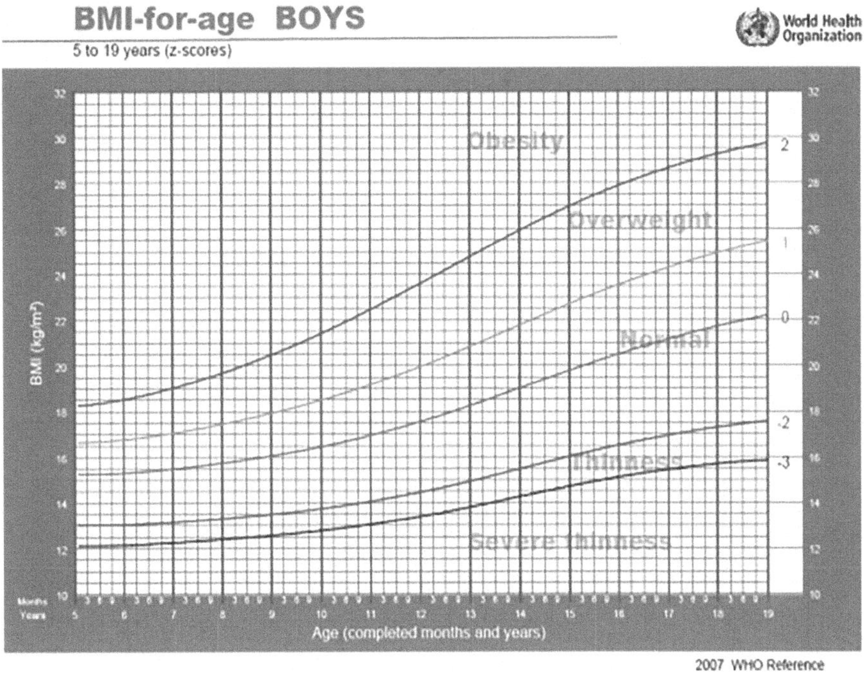


**Fig. 7.3** Example of WHO growth curves: BMI for age for boys. (Source: De Onis et al. 2007)

tall). For that reason, BMI-for-age charts are used to complement height for age for assessment and monitoring of thinness (low BMI for age), overweight and obesity (high BMI for age), and stunting (low height for age) in adolescents (WHO 1995; De Onis et al. 2007).

### 7.2.3 *Interpretation*

Once standardized methods and techniques have been applied and data recorded using appropriate tools, measurements are plotted on sex-specific growth charts (WHO 1995). In order to draw accurate interpretation, it is imperative to understand how to properly plot measurements and read data (Rogol et al. 2000; WHO 2008).

#### Plotting the Data

Figure 7.4 shows an example of a growth chart referring to important terms to understand before plotting. The horizontal line at the bottom of the graph is referred to as the *x-axis*. It shows either age (completed months and years) as in the figure or height (centimeters). The vertical line represents the *y-axis*. It shows either height

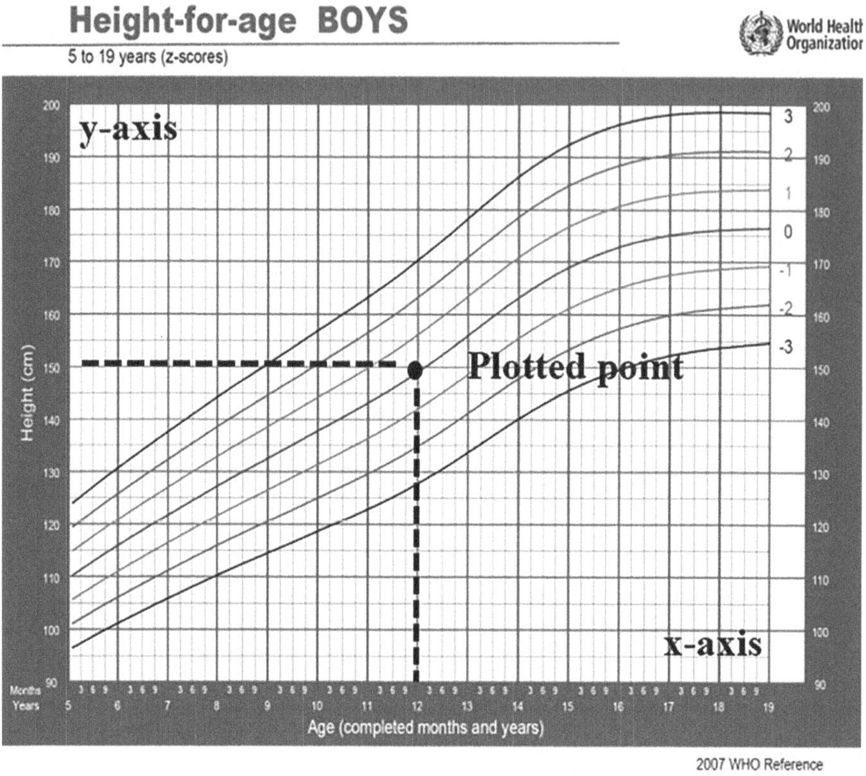


**Fig. 7.4** Example of height-for-age growth chart for boys. (Source: WHO 2008)

(cm) as in the figure, weight (kg), or BMI (kg/m$^2$). The point where the two axes meet is referred to as the *plotted point*. It is important to note that a single point is insufficient to make a diagnosis. Instead, multiple points are plotted over time to depict trends and determine if an adolescent's growth is normal or a deviation from the reference population (Rogol et al. 2000; WHO 2008).

## Interpreting Plotted Data

Curved lines on the chart represent results as either percentiles or z-scores (also known as standard deviation (SD) scores). They are different in terms of calculation but both appropriate for interpretation; however, the WHO recommends the use of z-scores (WHO 1995; De Onis et al. 1997; WHO 2008). Figure 7.5 demonstrates the systems' similarity. In the z-score system, the middle line labeled **(0)** is considered the median, as well as the line labeled **(50th)** in the percentile system. The median or 50th percentile designates the mean age of adolescents having the measured height, weight, or BMI in the reference population. The other lines represent deviation from average, and the further the plotted point is from the average in

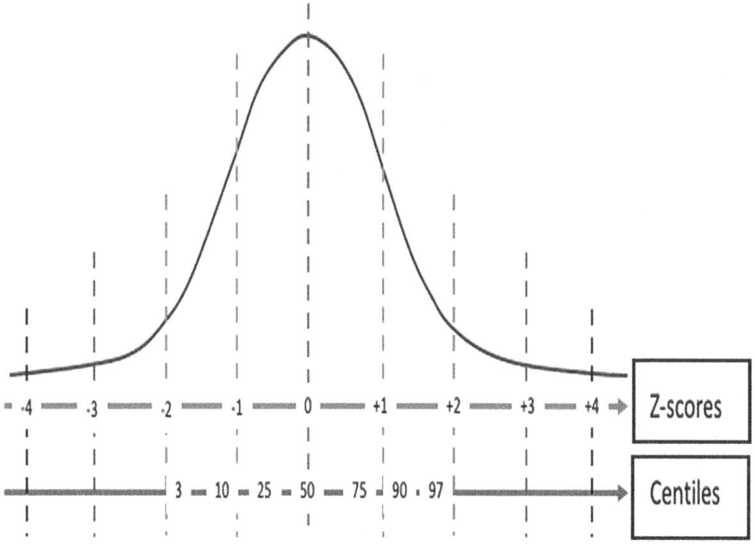


**Fig. 7.5** Z-score and percentile system equivalence. (Source: WHO 2008)

**Table 7.1** Cutoff points for height for age, weight for age, and BMI for age

| Z-score | Height-for-age indicators | Weight-for-age indicators | BMI-for-age indicators |
|---|---|---|---|
| Above 3 | Tallness | Better to use BMI for age | Obese |
| Above 2 | Normal | Better to use BMI for age | Obese |
| Above 1 | Normal | Better to use BMI for age | Overweight |
| 0 (median) | Normal | Normal | Normal |
| Below − 1 | Normal | Normal | Normal |
| Below − 2 | Stunted | Underweight | Wasting |
| Below − 3 | Severely stunted | Severely underweight | Severe wasting |

Source: (WHO 2008)

either direction (above or below), the greater the risk of growth problems (De Onis et al. 1997). Following the WHO recommendation, z-score system will be used in this chapter for interpretation.

## Cutoff Points

Cutoff points refer to the level or point which presents a limit to what is being studied. Thus, they are used for the purpose of expressing growth by means of z-scores (De Onis et al. 1997). For example, a plotted point between (−1, +1) SD or z-score line indicates normal growth. Moreover, a point that is $<-1$ SD and $\geq -2$ SD or $>+1$ SD and $\leq +2$ indicates risk (AKA preclinical level), and a point that is $<-2$ and $\geq +2$ SD indicates higher risk (AKA clinical level) (De Onis et al. 1997; WHO 2008).

The values presented in Table 7.1 shows cutoff points for height-for-age indicator. Stunting is indicated by low height for age with a cutoff value of <−2 SD. Weight should be closely monitored for adolescents with severe pathological stunting (>−3 SD) as it is associated with higher risk of overweight (Popkin et al. 1996). A z-score of >3 indicates tallness, which is less likely but if extreme may be related to an endocrine disorder and needs further investigation. With respect to weight-for-age indicator, a z-score of zero (median) to <-1 indicates normal weight, whereas a z-score of <-2 and <-3 indicates underweight and severely underweight, respectively. Cutoff points for BMI for age are also indicated in Table 7.1 (WHO 2008).

## 7.3 Results

Recent studies on adolescent health emphasize the significance of growth assessment for its integral role in monitoring health and nutritional status at individual and population levels (Christian and Smith 2018; Pimenta et al. 2011). Physical growth is widely recognized as a basic indicator to adolescent health. In the past, adolescence growth assessment was challenging due to the limited ability to differentiate between normal variations in growth from ones related to health risks (Christian and Smith 2018).

The 2007 WHO growth references for adolescents can be used to properly assess growth and address issues of malnutrition and growth problems based on indicators and cutoff points as mentioned beforehand. Normally growing adolescents usually follow the curves on the growth charts whether exactly on the line or in between lines. Alert is raised indicating an increased risk or a growth problem when the growth line crosses a z-score line or shows a sharp incline or decline or a flat line. Deviation from normal growth in adolescents translates into stunting, wasting, underweight, and overweight and obesity (De Onis et al. 1997).

### *7.3.1 Stunting*

Height for age is used as an indicator for stunting (short for age, <−2 SD), which may reflect undernutrition or illness. In adolescence, it reflects long-term underlying poor health, illness, and poor nutritional status that results in suboptimal linear growth. Moreover, environmental stress and poor socioeconomic conditions with higher risk for infection is associated with stunted growth (WHO 1995; Christian and Smith 2018). Non-pathological reasons for short stature include familial short stature, which is genetic in nature. The adolescent will present normal overall growth but may become short as an adult, given the growth history of the parents (Zeferino et al. 2003). Weight should be closely monitored for adolescents with severe pathological stunting (>−3 SD) as it is associated with higher risk of overweight (Popkin et al. 1996).

Limited studies in the MENA region investigated the prevalence of stunting among adolescents. In the Arabian Gulf, a study in Saudi Arabia measured the prevalence of stunting in adolescents (5–17 years) and compared results among three different references: the 2007 WHO charts, the 2000 Centers for Disease Control (CDC), and the 1978 National Center for Health Statistics (NCHS)/WHO references (El Mouzan Mohammad et al. 2011). Results indicated an increasing prevalence of stunting with age. Moreover, it concluded that the difference in the prevalence of short stature was not significant between any of the pairs of the three references (El Mouzan Mohammad et al. 2011).

As for Northern Africa, stunting was measured among a sample of Libyan school children (9–11 years) and assessed using the 2007 WHO charts. Results revealed a low prevalence of stunting compared to other countries in the region; however, the prevalence was higher in girls compared to boys (Elhisadi et al. 2013). The prevalence of thinness, stunting, and anemia using the 2007 WHO charts was investigated in a study among Sudanese children (6–14 years) and indicated acute undernutrition in about a quarter of the sample. Moreover, stunting was found to be more significantly common in children less than 10 years old (Mohamed and Hussein 2015). Similarly, another cross-sectional study in Sudan assessed the nutritional status among adolescent schoolgirls (11–18 years) and revealed a high prevalence of stunting and anemia among the study sample (Mahgoub et al. 2017). In Iraq, a study was conducted on schoolchildren aged 7–12 years in Baghdad to address stunting and measure its extent. Height for age was used to assess stunting in accordance with the WHO references. The study revealed stunting in one out of five of the children being slightly higher in girls and indicated an increasing height deficit with age (Al Saffar 2009).

### 7.3.2 Overweight and Obesity

Evidence indicates an alarming increase in overweight and obesity prevalence globally occurring in about one in four adolescents (WHO 2016). Several studies have associated obesity with higher risk for morbidity and mortality (Katanoda et al. 2019; Allcock et al. 2009). A study in Spain found that childhood obesity as well as underweight increases the risk for developing adult metabolic syndrome (Pimenta et al. 2011). Similarly, a study in Japan concluded that both underweight at adolescence and overweight are associated with higher risk of adult-onset diabetes (Katanoda et al. 2019).

In Yemen, obesity and overweight prevalence in primary school children aging 6–16 years was assessed. Standardized techniques were used to measure weight, height, and blood pressure, and cutoff values were used to identify nutritional status based on the WHO Expert Committee in Overweight and the 2007 WHO growth reference standards. The study revealed that one in five of the children was overweight and obese. Also, it found a significant association between blood pressure and excess body weight showing higher rates of pre-hypertensive children who are overweight than wasted or normal weight (Badi et al. 2012).

In Lebanon, a study investigated dietary, lifestyle, and socioeconomic correlates of overweight and obesity in children and adolescents 6–19 years. Overweight and obesity were classified according to the 2007 WHO growth references. Findings indicated a high prevalence of overweight and obesity in about one third of age group. The study concluded that results were comparable to rates in other countries in the region such as Bahrain and Syria, higher than rates in Qatar, and lower than those in the United Arab Emirates, and higher obesity rates were observed in boys than in girls (Nasreddine et al. 2014). In Jordan, the prevalence of obesity among schoolgirl adolescents was assessed. Anthropometric data was measured and compared to the 2007 WHO references. The study revealed that about a half of the subject were overweight and obese and addressed an alarming need for nutritional evaluation for this age group (Hamad et al. 2016).

Furthermore, a study in Lebanon compared trends in overweight and obesity in two national cross-sectional surveys (1997 and 2009). Overweight and obesity were defined using the new 2007 WHO references for adolescents. It highlighted a significant surge in obesity prevalence over a 12-year difference between the two studies (Nasreddine et al. 2012).

### *7.3.3 Wasting and Underweight*

Wasting or thinness is defined by a z-score of $<-2$ and is generally a result of recent and severe weight loss associated with undernutrition or protein-energy malnutrition due to starvation or chronic disease (De Onis et al. 1997). It provides a short-term indicator of poor nutritional status and increased risk of morbidity and mortality (Katanoda et al. 2019).

Undernutrition is one of the most common reasons of growth problems in adolescents, resulting either from starvation and poverty or from self-induced restrictions in dietary intake. Additionally, nutritional status affects sexual development; delayed onset age of menstruation is usually accompanied by undernutrition, whereas obesity is associated with early maturation (Rogol et al. 2000).

A concurrent existence of under- and over-nutrition was reported regarding the nutritional status of adolescents as moderate to high prevalence rates of stunting and overweight/obesity (Nasreddine et al. 2014). A high prevalence of thinness was observed in Sudan in over a third of the study group, and a less prevalence of underweight adolescents (10%) was found in Yemen (Badi et al. 2012; Mahgoub et al. 2017).

Nutritional demands increase in adolescence accompanying rapid growth. Thus, providing adequate energy and distribution of micro- and macronutrients play a role in proper growth. Malnutrition refers to the state of imbalance between what nutrients the body needs and the actual intake, leading to either undernutrition or overnutrition. This grants the term "double burden" of malnutrition (FAO 2006).

## 7.4 Discussion

Considering the fact that the highest linear growth characterizes the first 2–3 years of life and a similar high rate is observed during puberty, monitoring and evaluating growth is of utmost importance (Tanner and Whitehouse 1976). Nonetheless, very limited data are available on adolescents' growth within the Middle East and North Africa region. This highlights the need to focus on this age group due to a wide range of physiological and psychosocial changes they experience among a variety of factors playing a role in their overall growth and development.

Evidence suggests that due to this rapid growth, adolescence could provide a second opportunity for height catch-up growth in stunted children as height-for-age scores have been found to show improvement and possible recovery (Prentice et al. 2013). However, despite the fact that some catch-up growth with regard to height is possible, persisting issues in the MENA region, such as poor economic status and deprivation, should be taken into consideration as barriers to achieving optimal growth and development (Al Saffar 2009).

Increasing prevalence of overweight and obesity among adolescents characterizes it as one of the most challenging health issues of the century. Excess body weight during this phase is a risk factor for adulthood overweight and obesity. Moreover, it is associated with increased risk of hypertension, insulin resistance, and fatty liver disease among others that if left untreated persists through adulthood (Katanoda et al. 2019). Therefore, it is crucial to monitor body mass in the population on a regular basis using proper tools and cutoffs as excess body weight and increased BMI values can predict an inevitable rise in various health problems.

Assessing the nutritional status of adolescents and providing prevalence data for monitoring growth problems is of utmost importance on condition that cutoff values and reference charts are used in a consistent manner among healthcare establishments and researchers on local as well as international level. Nonetheless, no agreement on which cutoffs or reference data to use exists, making it challenging to compare prevalence studies, and in case of comparison interpretation, it should be dealt with caution (El Mouzan Mohammad et al. 2010).

The new 2007 WHO growth charts and references were developed stemming from the need to obtain a reliable screening tool for schoolchildren and adolescents worldwide. The WHO constructed the growth references by reconstructing the 1977 NCHS/WHO existing references from 5 to 19 years alongside data from the WHO Child Growth Standards to ensure a smooth transition between age groups. One emphasized advantage of these charts is that they allow further international reference among countries and populations (De Onis et al. 2007).

Very scarce research on growth assessment and prevalence of growth problems using the 2007 WHO exist. Despite it being a worldwide applicable tool, many healthcare establishments and researchers are still using the 2000 CDC reference. El Mouzan et al. investigated the prevalence of overweight and obesity in Saudi adolescents using the 2007 WHO references and compared the results to the 2000 CDC references. They observed that the prevalence of overweight and obesity in their sample using the 2007 WHO reference surpassed the one using the 2000 CDC reference. They clarified how this difference may be due to the different characteristics of the two references and the reference population. Moreover, the 2000 CDC charts underestimated the prevalence of overweight and obesity (El Mouzan Mohammad et al. 2010).

In accordance with the importance of growth assessment and proper standardization, the WHO has established the Global Action for Measurement of Adolescent health (GAMA) in collaboration with worldwide organizations including UNAIDS, UNESCO, UNFPA, UNICEF, UN Women, the World Bank Group, and the World Food Programme (WFP). It aims to improve and achieve consistent, standardized adolescent health measurement worldwide. Furthermore, it provides progress tracking toward better adolescent health (WHO 2018).

Investing in adolescent health is vital to a prospering future, and it all starts with proper assessment and interpretation of their growth as it is the building block for monitoring and identifying abnormalities. Growth monitoring is crucial as it provides information on adolescents' growth and development using simple tools and techniques, which should facilitate reinforcing global efforts that pour into overall adolescent health and well-being (Abul Fadl et al. 2010).

In conclusion, there is a clear need for more research into adolescent-specific growth parameters and assessment. Key issues to research include filling anthropometry data gaps and developing more health and growth indicators and shedding light on underrecognized growth problems such as stunting. Moreover, there has to be a consensus among healthcare institutions globally on which references to use. As discussed within the chapter, the 2007 WHO reference provides the most up-to-date and informed reference to monitor growth in adolescents and school-aged children and consequently is recommended for international use. Evidently, drawing inferences and comparing data from the literature is indeed challenging due to differences in sample selection, study age groups, and the type of references used.

If the countries were to use similar charts, it is imperative to provide frequent sufficient training sessions and monitoring to ensure standardized methods and techniques. Lastly, implementing intervention strategies toward healthy growth and decreasing risk of preventable growth problems have the potential to improve disease outcomes, reproductive health, and economic productivity in the population.

## Reflection Questions

1. If countries were to use standardized growth charts for assessment and monitoring, what other gaps need to be filled with regard to adolescent growth assessment?
2. The WHO growth charts represent standards for global population growth globally. Should countries develop local cutoff points based on local uses, growth patterns, and resources?

3. Proper measurement and interpretation require standardization of methods and techniques. In what ways can this be improved in the context of the MENA region?
4. For obtaining accurate interpretation, it is important to comprehend the plotting and reading of measurements. How important is it to educate mothers on this and how can that be achieved?
5. Nutritional problems are prevalent during adolescence due to rapid growth and increasing nutrient needs. What intervention strategies can be applied on the community level to prevent and reduce their occurrence?

## References

Abul Fadl, A., Bagchi, K. & Cheikh Ismail, L. 2010. Practices in child growth monitoring in the countries of the Eastern Mediterranean Region. *EMHJ-Eastern Mediterranean Health Journal, 16 (2), 194–201, 2010.*

Al Saffar, A. J. 2009. Stunting among primary-school children: a sample from Baghdad, Iraq.

Allcock, D. M., Gardner, M. J. & Sowers, J. R. 2009. Relation between Childhood Obesity and Adult Cardiovascular Risk. *International Journal of Pediatric Endocrinology,* 2009, 108187.

Azzopardi, P., Kennedy, E. & Patton, G. 2017. Data and indicators to measure adolescent health, social development and well-being. *Florence: UNICEF Office of Research, Innocenti.*

Badi, M. A. H., Triana, B. E. G. & Martinez, R. S. 2012. Overweight/obesity and hypertension in schoolchildren aged 6–16 years, Aden governorate, Yemen, 2009.

Cheikh Ismail, L., Knight, H. E., Ohuma, E. O., Hoch, L., Chumlea, W. C., For the International, F. & Newborn Growth Consortium for the 21st, C. 2013. Anthropometric standardisation and quality control protocols for the construction of new, international, fetal and newborn growth standards: the INTERGROWTH-21st Project. *BJOG: An International Journal of Obstetrics & Gynaecology,* 120, 48–55.

Christian, P. & Smith, E. R. 2018. Adolescent Undernutrition: Global Burden, Physiology, and Nutritional Risks. *Annals of Nutrition and Metabolism,* 72, 316–328.

De Onis, M., Blossner, M. & Organization, W. H. 1997. WHO global database on child growth and malnutrition. World Health Organization.

De Onis, M., Onyango, A. W., Borghi, E., Siyam, A., Nishida, C. & Siekmann, J. 2007. Development of a WHO growth reference for school-aged children and adolescents. *Bulletin of the World health Organization,* 85, 660–667.

De Onis, M., Onyango, A. W., Van Den Broeck, J., Chumlea, W. C. & Martorell, R. 2004. Measurement and standardization protocols for anthropometry used in the construction of a new international growth reference. *Food and nutrition bulletin,* 25, S27–S36.

El Mouzan Mohammad, I., Al Herbish Abdullah, S., Al Salloum Abdullah, A., Foster Peter, J., Al Omer Ahmad, A. & Qurachi Mansour, M. 2011. Prevalence of short stature in Saudi children and adolescents. *Annals of Saudi Medicine,* 31, 498–501.

El Mouzan Mohammad, I., Foster Peter, J., Al Herbish Abdullah, S., Al Salloum Abdullah, A., Al Omer Ahmad, A., Qurachi Mansour, M. & Kecojevic, T. 2010. Prevalence of overweight and obesity in Saudi children and adolescents. *Annals of Saudi Medicine,* 30, 203–208.

Elhisadi, T. A., Flailih, P. A. M., Almasmary, A. & Almansoory, W. K. 2013. Comparison of the Nutritional Status of Libyan School Children Aged 9–11 Years with the WHO-2007 Reference Population. *Science Journal of Medicine and Clinical Trials,* 2013.

Fao 2006. *The double burden of malnutrition: Case studies from six developing countries*, Food & Agriculture Org.

Gibson, R. S. 2005. *Principles of Nutritional Assessment,* New York, Oxford University Press.

Hamad, H. J., Abu-Hassouneh, D. T., Ibrahim, M. O., Al-Islam, M. E. & Faris, E. 2016. Prevalence of obesity among Jordanian school-aged adolescents in Greater Amman. *Prevalence,* 33, 91–96.

Katanoda, K., Noda, M., Goto, A., Mizunuma, H., Lee, J. S. & Hayashi, K. 2019. Being underweight in adolescence is independently associated with adult-onset diabetes among women: The Japan Nurses' Health Study. *Journal of Diabetes Investigation,* 10, 827–836.

Mahan, L. K. & Raymond, J. L. 2016. *Krause's Food & the Nutrition Care Process, Mea Edition E-Book*, Elsevier.

Mahgoub, H. M., Fadlelseed, O. E., Khamis, A. H., Bilal, J. A. & Adam, I. 2017. Nutritional and micronutrient status of adolescent schoolgirls in eastern Sudan: A cross-sectional study. *F1000Research,* 6.

Mohamed, S. & Hussein, M. D. 2015. Prevalence of Thinness, Stunting and Anemia Among Rural School-aged Sudanese Children: A Cross-sectional Study. *Journal of Tropical Pediatrics,* 61, 260–265.

Nasreddine, L., Naja, F., Akl, C., Chamieh, M. C., Karam, S., Sibai, A.-M. & Hwalla, N. 2014. Dietary, lifestyle and socio-economic correlates of overweight, obesity and central adiposity in Lebanese children and adolescents. *Nutrients,* 6, 1038–1062.

Nasreddine, L., Naja, F., Chamieh, M. C., Adra, N., Sibai, A.-M. & Hwalla, N. 2012. Trends in overweight and obesity in Lebanon: evidence from two national cross-sectional surveys (1997 and 2009). *BMC Public Health,* 12, 798.

Pimenta, A. M., Beunza, J.-J., Sanchez-Villegas, A., Bes-Rastrollo, M. & Martinez-Gonzalez, M. A. 2011. Childhood underweight, weight gain during childhood to adolescence/young adulthood and incidence of adult metabolic syndrome in the SUN (Seguimiento Universidad de Navarra) Project. *Public Health Nutrition,* 14, 1237–1244.

Popkin, B. M., Richards, M. K. & Montiero, C. A. 1996. Stunting is Associated with Overweight in Children of Four Nations That Are Undergoing the Nutrition Transition. *The Journal of Nutrition,* 126, 3009–3016.

Prentice, A. M., Ward, K. A., Goldberg, G. R., Jarjou, L. M., Moore, S. E., Fulford, A. J. & Prentice, A. 2013. Critical windows for nutritional interventions against stunting. *The American Journal of Clinical Nutrition,* 97, 911–918.

Rogol, A. D., Clark, P. A. & Roemmich, J. N. 2000. Growth and pubertal development in children and adolescents: effects of diet and physical activity. *The American Journal of Clinical Nutrition,* 72, 521S–528S.

Sawyer, S. M., Azzopardi, P. S., Wickremarathne, D., & Patton, G. C. 2018. The age ofadolescence. *The Lancet Child & Adolescent Health,* 2(3), 223–228. https://doi.org/10.1016/S2352-4642(18)30022-1

Tanner, J. M. & Whitehouse, R. H. 1976. Clinical longitudinal standards for height, weight, height velocity, weight velocity, and stages of puberty. *Archives of Disease in Childhood,* 51, 170.

WHO 1995. Physical Status: The Use and Interpretation of Anthropometry. *WHO Technical Report Series.* Geneva, Switzerland: World Health Organization.

WHO 2008. WHO child growth standards: training course on child growth assessment.

WHO 2016. Overweight and obesity. Retrieved 15 February 2022 from, https://www.who.int/gho/ncd/risk_factors/overweight_obesity/obesity_adolescents/en/

WHO 2018. About GAMA. Retrieved 15 February 2022 from https://www.who.int/data/maternal-newborn-childadolescent/gama/about-gama

Zeferino, A., Barros Filho, A. A., Bettiol, H. & Barbieri, M. A. 2003. Monitoring growth. *Jornal de Pediatria,* 79, S23–S32.

# Chapter 8
# Oral Health of Adolescents in the MENA Region

**Reham Khaled Abou El Fadl and Haneen Raafat Fathi Mousa**

## 8.1 Introduction

Globally, around 1.2 billion people are aged 10–19 years, totaling 16% of the world population of which 75 million live in the Middle East and North Africa (MENA) (UNICEF 2019). According to the life course approach to health—which advocates for starting disease prevention at early-life stages through to adulthood (Jacob et al. 2017)—adolescence is considered a sensitive period of rapid development, experimentation, and choice-making during which adolescents establish certain behavioral patterns. Adopting unhealthy behaviors such as poor diet and high sugar intake, alcohol consumption, substance use, smoking, and the use of smokeless tobacco during this phase might contribute to lifetime detrimental health consequences in adulthood and even jeopardize the health of the next generations (WHO 2018). Most of those health-comprising behaviors are common risk factors between oral

The world's Middle-Income Countries (MICs) are a diverse group by size, population, and income level. They are defined as lower middle-income economies (such as Egypt Arab Republic, Morocco & Tunisia those with a gross national income (GNI) per capita between $1036 and $4045; and upper middle-income economies (such as Jordan, Iraq, and Libya) – those with a GNI per capita between $4046 and $12,535 (World Bank Country and Lending Groups: https://datahelpdesk.worldbank.org/knowledgebase/articles/906519-world-bank-country-and-lending-groups).

R. K. Abou El Fadl (✉) · H. R. F. Mousa
Faculty of Dentistry, Ain Shams University, Cairo, Egypt
e-mail: rehamabouelfadl@dent.asu.edu.eg; haneenraafat@dent.asu.edu.eg



C. Barakat et al. (eds.), *Adolescent Health in the Middle East and North Africa*, Global Perspectives on Health Geography,
https://doi.org/10.1007/978-3-030-92107-1_8

diseases and other chronic conditions such as obesity, diabetes, cardiac problems, and some cancers (Sheiham and Watt 2000). Though, in general, oral health is considered a reliable indicator of an individual's overall health and well-being, minimal attention is paid to promoting adolescents' oral health in both practice and policy (Percy 2008). As a consequence, globally, the prevalence of unmet dental needs is quite high in this age group, with highest rates in the Eastern Mediterranean region (84.2%) and Africa (78.0%) (Ghafari et al. 2019). At least three out of every ten adolescents between 13 and 18 years experience dental pain at some point in life. (Pentapati et al. 2020). Moreover, of all health services, oral care is usually the most commonly unmet health need among children and adolescents with special healthcare needs (SHCN) (Lewis 2009). Access to dental care usually poses a challenge for people with SHCN due to the inability to afford treatment costs or difficulty finding a dental practitioner who is willing and capable to treat this population (Norwood and Slayton 2013).

## 8.2 Methodology

A comprehensive literature review was conducted to explore the common oral health problems in adolescence and their prevalence in the MENA region. Two electronic databases—*Web of Science* and *Embase*—were searched, and only English articles were assessed. The review included research studies based in countries in the MENA region, particularly in the last 5 years (2015–2020). Editorials and commentaries were excluded.

The search strategy included the following combination of controlled vocabulary and keywords: [Adolescen* OR teen* OR preteen* OR "high school students" OR Underage] AND ["Oral health" OR "dental health" OR "dental disease" OR "oral disease" OR dental caries OR "DMFS" OR "DMFT" OR "defs" OR "deft" OR "Cast index" OR"ICDAS" OR "periodontitis" OR "periodontal index" OR "periodontal pockets" OR gingivitis OR "gingival index" OR "dental plaque" OR "oral hygiene" OR "dental hygiene" OR flossing OR toothbrushing OR "orthodontic needs" OR "oral lesions" OR "Oral cancer" OR "dental trauma" OR "oro-facial trauma" OR "oral habits" OR "dental fluorosis" OR "dental neglect" OR "dental needs" OR "dental pain" OR "oral pain"].

All the references were imported to EndNote, and duplicates were identified by EndNote and subsequently removed. Titles and abstracts of all entries were analyzed by two reviewers, and only full texts of relevant references were retrieved and reevaluated. Additionally, the Global Burden of Disease (GBD) 2017 dataset was used to generate epidemiologic data on oral conditions among adolescents and youth in the MENA region. In order to acquire data relevant to adolescence and teenage, only the 10–14 and 15–19 age groups were selected. Graphs depicting disease frequencies were developed using the *GBD Results Tool*.

## 8.3 Findings

### *8.3.1 Common Oral Health Risks in Adolescents*

#### Dental Caries

Adolescence marks a period of increased activity of dental caries (tooth decay) due to the immaturity of the recently erupted permanent teeth, increased intake of sugary diet and beverages, and lack of attention to oral hygiene practices in this age group ("Adolescent Oral Health Care" 2017). When left untreated, carious lesions lead to pain and infection, which could impair the child/adolescent's ability to eat and sleep, thus resulting in poor growth and development. Tooth decay has been strongly associated with school absenteeism and low academic performance in adolescents, and parents have reported experiencing distress and sense of guilt, work absenteeism, and financial hardships due to their children's poor oral health conditions. In some instances, owing to the increased esthetic desire and awareness in this age group, when multiple front teeth are involved, adolescents might refrain from smiling or interacting with others due to compromised esthetics, which in turn reduce their self-esteem and impair their social relationships ("Adolescent Oral Health Care" 2017).

Though preventable, globally, the burden of dental caries is still high and varies considerably both between and within countries. Middle-income countries (MICs)* have the highest burden of dental caries among 12-year-old adolescents, two thirds of which are left untreated. Even in high-income countries, more than half of tooth decay, in this age group, is not managed (FDI World Dental Federation 2015), which unveils the high rates of unmet dental needs in adolescence worldwide.

In the MENA region, however, data sources on trends in dental caries in adolescence per se are limited in most countries. One national study in Kuwait reported a 14% prevalence of caries among 12-year-olds (Alqaderi et al. 2016), while according to Egypt's 2013–2014 national oral health survey, around 70% of both children and adolescents aged 4–18 years experienced some form of tooth decay (WHO. EMRO 2014). In Saudi Arabia, the prevalence rates of dental caries in adolescents were determined at the province level and varied from over 90% in Jazan to 64.98% in the Riyadh region. (Al-Rafee et al. 2019; Aljanakh 2017; Alshahrani et al. 2018; Alzahrani 2018; Quadri et al. 2015). Figure 8.1 presents the prevalence rates of dental caries among adolescents aged 10–14 and 15–19 years in the MENA region in comparison with global statistics based on findings from the GBD study of 2017.

Noticeably, in the MENA countries, there has been a decline in the prevalence of dental caries among 10–14- and 15–19-year-olds. As revealed in Fig. 8.1, gender differences seem to be consistent at both global and regional levels, where female adolescents tend to be at higher risk to develop tooth decay than males in both age groups. On the other hand, unlike global data estimates, in the MENA countries, younger adolescents (10–14 years) in both genders had higher caries prevalence rates than those aged 15–19 years (Fig. 8.1).

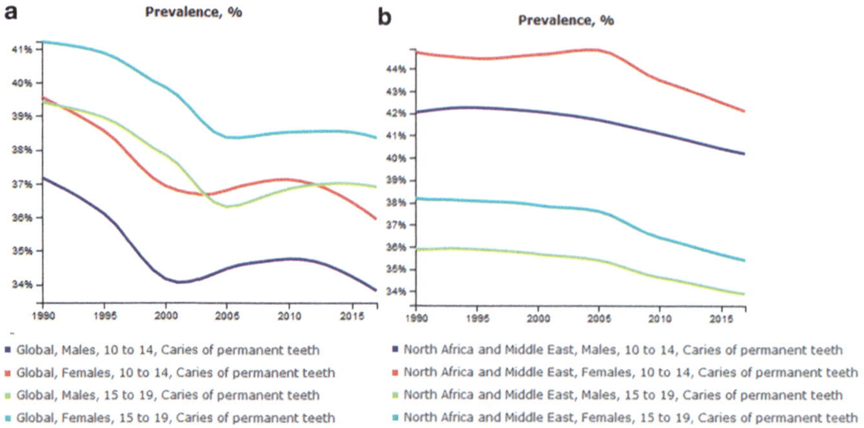


**Fig. 8.1** (**a**) Global prevalence of dental caries among adolescents aged 10–14 and 15–19 years (Source: Institute for Health Metrics Evaluation. Used with permission. All rights reserved). (**b**) Prevalence of dental caries among adolescents aged 10–14 and 15–19 years from the MENA. (Source: Institute for Health Metrics Evaluation. Used with permission. All rights reserved)

## Periodontal Diseases

Periodontal diseases refer collectively to a myriad of conditions that affect the gums, the supporting dental tissues, and/or the alveolar bone. Gingivitis (which is a chronic reversible inflammation of the gums) is a very common condition in adolescents. The prevalence of gingivitis in adolescents is higher than in adults and younger children ("Adolescent Oral Health Care" 2017), mostly due to hormonal fluctuations during puberty. Sex hormones might cause gingivitis through altering the composition of oral microflora, affecting fluid accumulation in the gingival tissues, or modifying the inflammatory response of the gingival tissues ("Adolescent Oral Health Care" 2017; Armitage 1986). Gingivitis in adolescents could also be secondary to mouth breathing, orthodontic therapy, pregnancy, smoking, or intake of certain drugs ("Adolescent Oral Health Care" 2017; Al-Ghutaimel et al. 2014).

Another advanced form of gum disease is periodontitis, a serious condition that affects the tooth-supporting tissues and the alveolar bone (FDI World Dental Federation 2015). All forms of periodontitis can occur during adolescence; however, aggressive (juvenile) periodontitis is more common during the circumpubertal period, while chronic periodontitis is only seen occasionally (Al-Ghutaimel et al. 2014; Armitage 1986).

Estimates of the frequency of gingival and periodontal problems among adolescents are available from very few countries in the MENA region. A nationally representative study of Kuwait, in 2016, revealed that 42.5% of 12-year-old Kuwaiti adolescents had gingivitis (Alqaderi et al. 2016), while in Saudi Arabia, the prevalence was as high as 65% in a nationally representative sample of high school-attending adolescents (AlGhamdi et al. 2020b). On the other hand, according to

(AlGhamdi et al. 2020a), the prevalence of periodontitis among 15–19-year-olds in Saudi Arabia was 8.6%, while in one study in Benslimane Province in Morocco, it was found that 3.7% and 2.6% of adolescents aged 12–15 and 4.9% and 7% of adolescents aged 16–18 had aggressive and chronic periodontitis, respectively (Kissa et al. 2016).

Findings from the GBD data (up to 2017) have shown that significant variations exist between global and regional frequencies of periodontal diseases among adolescents (Fig. 8.2). This could be attributed to the paucity of up-to-date reliable evidence on disease burden in adolescence at the population level in most of the MENA countries. Another conspicuous difference between the MENA countries and the global data is the distinct gap in gender prevalence at the regional level.

As noticed in Fig. 8.2 male adolescents in the MENA countries have significantly higher burden of periodontal diseases which might possibly indicate common unhealthy practices such as smoking, khat chewing, or substance abuse.

## Malocclusion

Malocclusion refers to any abnormal position of the teeth or jaws, which leads to impaired orofacial functions and in severe cases facial asymmetry. The condition is usually associated with a combination of genetic and environmental factors and requires orthodontic and/or surgical intervention. Though not considered a pathological disease, malocclusion is a common problem in adolescent populations, which when overlooked could lead to serious consequences.

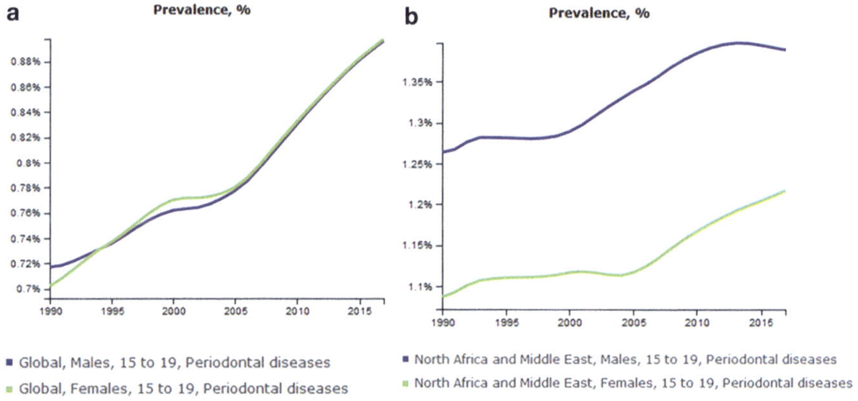


**Fig. 8.2** (**a**) Global prevalence of periodontal diseases among 15–19-year-olds (Source: Institute for Health Metrics Evaluation. Used with permission. All rights reserved). (**b**) Prevalence of periodontal diseases among 15–19-year-olds in the MENA countries. (Source: Institute for Health Metrics Evaluation. Used with permission. All rights reserved)

Among adolescents from different cultural backgrounds, malocclusion is known to adversely affect their psychological well-being and lower their self-esteem due to compromised facial esthetics (Dimberg et al. 2015).

Aside from the speech difficulties, malalignment of teeth could also increase the risk of sustaining dental trauma, particularly when upper anterior (front) teeth are protruded ("Adolescent Oral Health Care" 2017). Owing to pain, loss of function, and poor esthetics, traumatic injuries could be distressing at the physical, psychological, and emotional levels (Lee and Divaris 2009). An abnormal teeth arrangement or relation could also be associated with temporomandibular joint affection ("Adolescent Oral Health Care" 2017).

Furthermore, crowded teeth, being more difficult to cleanse, are more prone to food accumulation and plaque retention, which in turn increases the risk of dental caries and gingivitis (Kolawole and Folayan 2019). Based on one recent systematic review, it was found that the prevalence of unmet orthodontic treatment needs among adolescents in the MENA countries is as high as 40.8% (Ghafari et al. 2019), which highlights the necessity of directing more attention toward ensuring equitable access to orthodontic care in this age group.

## Orofacial Traumatic Injuries

Trauma to oral tissues could result in fracture, loss, or dislocation of teeth, bruises or laceration of the soft tissues, or fractures of the facial bones. Craniofacial trauma causes more than four million deaths worldwide, which is half of all the trauma-related deaths (FDI World Dental Federation 2015). Several risk factors for traumatic dental injuries among adolescents were identified including protrusion of anterior teeth, peer relationships, exposure to violence or abuse, as well as the socioeconomic level (Glendor 2009). Another prominent cause of oral injuries among adolescents is sports activities, during which they get exposed to facial injuries due to falls or collisions with other players or surfaces ("Adolescent Oral Health Care" 2017). Such injuries could have serious physical, psychosocial, and economic impacts on adolescents and their families.

Though teen years are a high-risk time for traumatic orofacial injuries, in the MENA region, only data on exposure of adolescents to serious injuries in general without reference to dental injuries is available. One study, however, in Jordan revealed that the prevalence of traumatic dental injuries was 14.6% among 12-year-old schoolchildren of Amman, indicating that one out of every seven adolescents has suffered from some form of dental trauma by the age of 12 (Rajab and Abu Al Huda 2019).

## Dental Fluorosis

Fluoride is considered among the most commonly studied protective factors of dental decay, despite that excess amounts of fluoride (above 1.5 mg/L) uptake can be destructive to permanent dentition. Dental fluorosis usually manifests as whitish or brownish discoloration or (in its severe form) pitting of teeth surface(s). Such defects start to be visible as soon as the permanent teeth erupt during childhood and early adolescence. It is also noteworthy that dental fluorosis is not only an esthetic problem, but it also increases teeth liability to decay.

Excessive fluoride content in drinking water is one of the major causes of dental fluorosis in some countries of the MENA region. Fluoride concentrations exceeding 10 mg/L were reported in countries such as Oman and Yemen (Aldeghaither 2018). One study in Yemen reported that 30.8% of 14-year-old Yemeni adolescents suffered from dental fluorosis (Kadir and Al-Maqtari 2010), and according to a national oral health survey in the UAE, up to 61% of 12-year-olds and 53% of 15-year-olds, in some areas in the country, had some form of clinical dental fluorosis (El-Nadeef et al. 2009).

## 8.4 Discussion and Future Implications

Adolescents should be regarded as a separate, unique population with special oral healthcare needs that differ substantially from those of adults or younger children, taking into consideration the following factors: (i) the potentially higher liability during this age group to various oral diseases like tooth decay, gingivitis, and oral trauma; (ii) the suboptimal dietary habits of adolescents that are affected by many social and environmental factors; (iii) the likelihood of adolescents to engage in risky behaviors like smoking, drug abuse, or alcohol consumption, which reflects on their oral and general health; (iv) some conditions like orthodontic problems or congenitally missing teeth start to be discernable in this age; (v) adolescents have special psychological needs that can affect dental health both directly and indirectly; (vi) the higher demand for esthetics particularly among adolescent girls; (vii) tendency to seek dental treatment might be hindered by dental anxiety in this age group; and (viii) some adolescents are also commonly affected by some other coexisting conditions that might further affect their oral health such as pregnancy, eating disorders, or systemic diseases ("Adolescent Oral Health Care" 2017).

Despite that, in most countries of the MENA region, the oral health of adolescents is a neglected area of research and reliable, up-to-date data on frequency and severity of different oral conditions, the psychosocial attributes of oral diseases, and the prevalence of unmet dental care needs in this age group are still lacking. Furthermore, oral health needs among adolescents and youth with special healthcare needs such as Down's syndrome, cerebral palsy, autistic disorder, and other genetic and hereditary disorders are not adequately studied in any of the region's countries. The absence of a comprehensive data collection system and the inadequacy of local research on oral health outcomes in adolescence and teenage hinder policy-making and limit the development of effective public health responses to promote adolescents' oral health at the country level.

Given the complex nature of oral health issues arising during adolescence, all regional countries need to develop a research agenda for surveillance of oral health outcomes and common health determinants influencing oral health-related quality of life in both early and late phases of adolescence. Addressing knowledge gaps in the identification of causal factors and management of oral diseases among adolescents at the country level and establishing national quality standards for the provision of oral health services in this age group are warranted. There is also a pressing need to engage the youth of diverse ages and social backgrounds, especially those who are most vulnerable such as youth with SHCNs, in advocacy programs, community assessment activities, and decision-making to secure the success of the policy-making process. Furthermore, non-dental healthcare providers should encourage and assist in the referral of adolescents for routine dental checkups and risk assessment.

## Reflection Questions

- How do adolescents' oral health needs differ from those of younger children?
- Why are oro-dental problems such as dental caries, gingival problems, and malocclusion highly prevalent among adolescents in the MENA region?
- How can oral diseases impact social and psychological development of adolescents?
- What actions need to be undertaken, at both national and regional levels, to promote adolescents' oral health in the MENA countries?
- How can promotion of adolescents' oral health be an effective tool in reducing global burden of oral diseases?

## References

Adolescent Oral Health Care, 2017. Pediatr. Dent. 39, 213–220.
Al-Ghutaimel, H., Riba, H., Al-Kahtani, S., Al-Duhaimi, S., 2014. Common periodontal diseases of children and adolescents. Int. J. Dent.
Al-Rafee, M.A., Al Shammery, A.R., Al Rumikan, A.S., Pani, S.C., 2019. A comparison of dental caries in urban and rural children of the Riyadh region of Saudi Arabia. Front. Public Heal.
Aldeghaither, 2018. Fluoride in the Greater Middle East and Its Source: A Literature Review.
AlGhamdi, A., Almarghlani, A., Alyafi, R., Ibraheem, W., Assaggaf, M., Howait, M., Alsofi, L., Banjar, A., Al-Zahrani, M., Kayal, R., 2020a. Prevalence of periodontitis in high school children in Saudi Arabia: a national study. Ann. Saudi Med.
AlGhamdi, A., Almarghlani, A., Alyafi, R., Kayal, R., Al- Zahrani, M., 2020b. Gingival health and oral hygiene practices among high school children in Saudi Arabia. Ann. Saudi Med.
Aljanakh, M., 2017. Prevalence and severity of dental caries among public school students aged 16-18 in Hai'l, Kingdom of Saudi Arabia. Int. J. Health Sci. (Qassim). 11, 50–53.
Alqaderi, H., Tavares, M., Hartman, M., Goodson, J.M., 2016. Effect of sleep and salivary glucose on gingivitis in children. J. Dent. Res.
Alshahrani, I., Tikare, S., Meer, Z., Mustafa, A., Abdulwahab, M., Sadatullah, S., 2018. Prevalence of dental caries among male students aged 15–17 years in southern Asir, Saudi Arabia. Saudi Dent. J.

Alzahrani, A.A., 2018. The Oral Health Status and Attitudes of Saudi Adolescent Male Students in Albaha Province: A Cross-Sectional Study. Adv. Dent. Oral Heal.

Armitage, G.C., 1986. Periodontal diseases of children and adolescents. CDA J. 14, 57–61.

Dimberg, L., Arnrup, K., Bondemark, L., 2015. The impact of malocclusion on the quality of life among children and adolescents: A systematic review of quantitative studies. Eur. J. Orthod.

El-Nadeef, M.A.I., Al Hussani, E., Hassab, H., Arab, I.A., 2009. National survey of the oral health of 12- and 15-year-old schoolchildren in the United Arab Emirates. East. Mediterr. Heal. J.

FDI World Dental Federation., 2015. The Challenge of Oral Disease–A Call for Global Action. The Oral Health Atlas, 2nd edition.

Ghafari, M., Bahadivand-Chegini, S., Nadi, T., Doosti-Irani, A., 2019. The global prevalence of dental healthcare needs and unmet dental needs among adolescents: a systematic review and meta-analysis. Epidemiol. Health.

Glendor, U., 2009. Aetiology and risk factors related to traumatic dental injuries - A review of the literature. Dent. Traumatol.

Jacob, C.M., Baird, J., Barker, M., Cooper, C., Hanson, M., 2017. The Importance of a Life Course Approach to Health: Chronic Disease Risk from Preconception through Adolescence and Adulthood, Who Report.

Kadir, R.A., Al-Maqtari, R.A.S., 2010. Endemic fluorosis among 14-Year-old Yemeni adolescents: An exploratory survey. Int. Dent. J.

Kissa, J., Chemlali, S., El Houari, B., Amine, K., Khlil, N., Mikou, S., Nadifi, S., Albandar, J.M., 2016. Aggressive and chronic periodontitis in a population of Moroccan school students. J. Clin. Periodontol.

Kolawole, K.A., Folayan, M.O., 2019. Association between malocclusion, caries and oral hygiene in children 6 to 12 years old resident in suburban Nigeria. BMC Oral Health.

Lee, J.Y., Divaris, K., 2009. Hidden consequences of dental trauma: The social and psychological effects. In: Pediatric Dentistry.

Lewis, C.W., 2009. Dental Care and Children with Special Health Care Needs: A Population-Based Perspective. Acad. Pediatr.

Norwood, K.W., Slayton, R.L., 2013. Oral health care for children with developmental disabilities. Pediatrics.

Pentapati, K.C., Yeturu, S.K., Siddiq, H., 2020. Global and regional estimates of dental pain among children and adolescents—systematic review and meta-analysis. Eur. Arch. Paediatr. Dent.

Percy, M.S., 2008. Oral health of adolescents--it's more than dental caries. MCN. Am. J. Matern. Child Nurs.

Quadri, F.A., Hendriyani, H., Pramono, A., Jafer, M., 2015. Knowledge, attitudes and practices of sweet food and beverage consumption and its association with dental caries among schoolchildren in Jazan, Saudi Arabia. East. Mediterr. Heal. J.

Rajab, L.D., Abu Al Huda, D., 2019. Impact of treated and untreated traumatic dental injuries on oral health-related quality of life among 12-year-old schoolchildren in Amman. Dent. Traumatol.

Sheiham, A., Watt, R.G., 2000. The Common Risk Factor Approach: A rational basis for promoting oral health. Community Dent. Oral Epidemiol.

UNICEF, 2019. Adolescents overview [WWW Document]. URL https://data.unicef.org/topic/adolescents/overview/

WHO.EMRO, 2014. Egypt releases results of epidemiological study on oral health status [WWW Document]. URL http://www.emro.who.int/egy/egypt-events/results-of-epidemiological-study-on-oral-health-status-released.html

WHO, 2018. Adolescents: health risks and solutions Factsheet [WWW Document]. URL https://www.who.int/news-room/fact-sheets/detail/adolescents-health-risks-and-solutions

# Chapter 9
# Demographic and Geographic Variations in Respiratory and Allergic Conditions Among Adolescents in the United Arab Emirates

**Rania Dghaim, Shatha Muhsineh, Fatma Nasser, and Caroline Barakat**

## Abbreviations

| | |
|---|---|
| AD | Atopic Dermatitis |
| AR | Allergic Rhinitis |
| COPD | Chronic Obstructive Pulmonary Disease |
| CRD | Chronic Respiratory Disease |
| DALY | Disability-Adjusted Life Years |
| ECRHS | European Community Respiratory Health Survey |
| GBD | Global Burden of Disease |
| GCC | Gulf Cooperation Council |
| ISAAC | International Study of Asthma and Allergies in Childhood |
| MOE | Ministry of Education |
| MOH | Ministry of Health |
| NCD | Non-communicable Disease |
| NSPHUAE | National Study of Population Health in the United Arab Emirates |
| RAK | Ras Al Khaimah |
| SPSS | Statistical Package for Social Sciences |
| UAE | United Arab Emirates |
| UAQ | Umm Al Quwain |
| US | United States |

R. Dghaim (✉) · F. Nasser
College of Natural and Health Sciences, Zayed University, Dubai, United Arab Emirates
e-mail: Rania.dghaim@zu.ac.ae

S. Muhsineh
Health Policies and Standards Department, Health Regulation Sector, Dubai Health Authority, Dubai, United Arab Emirates

C. Barakat
Faculty of Health Sciences, Ontario Tech University, Oshawa, ON, Canada
e-mail: Caroline.Barakat@ontariotechu.ca



C. Barakat et al. (eds.), *Adolescent Health in the Middle East and North Africa*,
Global Perspectives on Health Geography,
https://doi.org/10.1007/978-3-030-92107-1_9

USA United States of America
WHO World Health Organization

## 9.1 Introduction

Chronic respiratory diseases (CRDs) comprise a range of pulmonary disorders. The most prevalent CRDs include asthma and respiratory allergies, chronic obstructive pulmonary disease (COPD), allergic rhinitis or “hay fever,” sleep apnea, and pulmonary hypertension (WHO 2007). These affect millions worldwide (WHO 2019). The WHO Global Burden of Disease (2016) estimates that 239 million individuals suffer from asthma globally (GBD 2016 Disease and Injury Incidence and Prevalence Collaborators 2017; Razzak et al. 2020). Asthma is the most prevalent CRD, affecting all ages, genders, and ethnicities. While the prevalence of asthma and other respiratory allergies has increased steadily in recent decades, there are wide regional and geographical variations in prevalence, severity, and mortality between and within countries (Hussain et al. 2018).

For Western countries, a wealth of information is available on time trends and regional variations in asthma prevalence, whereas only scare and fragmented data is available on the Arab Gulf in general and the United Arab Emirates (UAE) in particular (Alsowaidi et al. 2010). The UAE is a federation of seven emirates (Abu

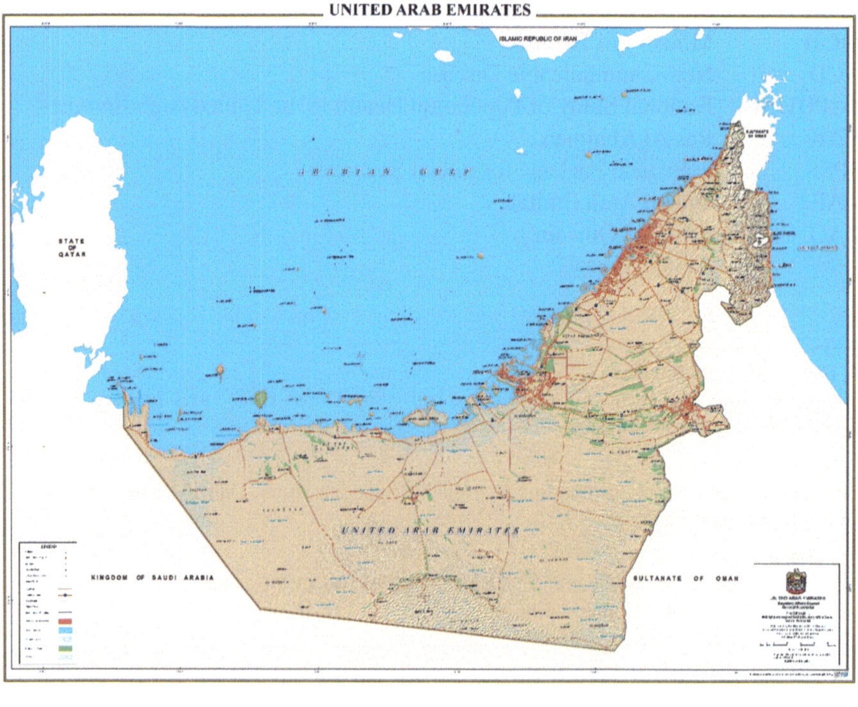


Fig. 9.1 Official map of the UAE (https://u.ae/-/media/)

Dhabi, Dubai, Sharjah, Ajman, Al Fujairah, Ras al Khaimah, and Umm Al Quwain (UAQ)). They all overlook the Arabian Gulf except for Al Fujairah, which overlooks the Gulf of Oman (Fig. 9.1). The UAE's population has recently increased dramatically, from 4.1 million in 2005 to 9.5 million in 2018 (Worldometers.info. 2019). This was caused by the significant growth in the UAE's economic and service sectors, which led to an influx of migrant workers and their families from various cultural and heritage backgrounds. UAE citizens comprise only about 11.5% of the population. The UAE's population is young, with the majority falling in the working age group of 25–54 years old (65.9%). Only 28% of the UAE's population are female, mainly due to the high proportion of male migrant workers. Children (younger than 14 years old) comprise 14.2% of the population, while adolescents and young adults (15–24 years) comprise 12.7% (GMI Blogger 2019).

Non-communicable diseases (NCDs) are a major public health challenge in the UAE as a prominent cause of death and disability, accounting for nearly 77% of all deaths and 17% probability of premature deaths (before the age of 70) in 2016 (Fadhil et al. 2019). Among the four major NCDs, chronic respiratory diseases account for 5% of all deaths in the UAE (WHO 2018).

### 9.1.1 Epidemiology of Respiratory Conditions and Allergies

Asthma is a chronic inflammatory disease characterized by bronchial hyperresponsiveness, recurrent attacks of wheeze, shortness of breath, chest tightness and pain, and nocturnal cough (Alzaabi et al. 2018; Al Ghobain et al. 2012; ISAAC 1993). Asthma is a major public health challenge for countries of all income levels. While asthma prevalence is higher in high-income countries, most asthma-related mortality occurs in low- to middle-income countries (WHO 2017). Asthma morbidity imposes a substantial burden on healthcare systems while reducing the quality of life of affected individuals and their families (Mahboub et al. 2010).

Although asthma affects all age groups, its incidence and prevalence are higher in children (ISAAC 1993). The International Study of Asthma and Allergies in Childhood (ISAAC) (ISAAC 1993) has provided information regarding the prevalence of asthma in children worldwide. Phase III of the ISAAC (conducted between 2000 and 2003) showed a global prevalence of 11.6% for wheeze (a good indicator of asthma diagnosis) for the age group of 6 and 7 years old and 13.7% for the age group of 13 and 14 years old. The highest childhood prevalence is in English-speaking and Western countries and Latin America. Rates are lower in Africa, the Indian Subcontinent, and the Eastern Mediterranean (Pearce et al. 2007; Bonamonte et al. 2019).

A 1992–1993 cross-sectional study of 850 school children living in both urban and rural areas in the UAE reported a relatively high prevalence of diagnosed asthma (13.6%), breathlessness, or chest tightness (9.7%) and nocturnal cough (8.9%) (Benner et al. 1994). In 2000, a study using the ISAAC questionnaire administered to 2300 children aged 6–13 years across the UAE's seven emirates reported similar rates of physician-diagnosed asthma (13%) and wheeze (15.6%) (al-Maskari et al. 2000). In a 2012 population study that assessed the prevalence of asthma in UAE adults (20–44 years) using the Standard European Community Respiratory Health

Survey (ECRHS), asthma prevalence of 9.8% was reported, with 10.1% of participants under the age of 20 reported a wheeze without cold and 13.5% were woken up by cough attacks (Mahboub et al. 2012).

Allergic rhinitis (AR), also known as hay fever, is a common allergic disorder caused by inflammation of the nasal mucosa by inhaled allergens. Common symptoms include sneezing, runny nose (rhinorrhea), and/or nasal congestion, in conjunction with conjunctivitis (Mahboub et al. 2014). According to the WHO (2007), over 400 million people suffer from AR worldwide. The average incidence among school-aged children ranges from 0.5% to 28% across countries. The prevalence of AR and other allergic respiratory diseases is increasing, especially in children, to reach 40% in some developed countries, although AR remains largely underdiagnosed and undertreated (Ann et al. 2015).

In 1993, AR was the most frequently seen respiratory illness among UAE school-aged children (6–14 years) at 22.9% (Bener et al. 1994), while a 2009 study conducted in Al-Ain City (Abu Dhabi Emirate) reported a prevalence of 36% for AR, using a modified form of the ISAAC questionnaire (Mahboub et al. 2014). A more recent study used the ECRHS-II to report a prevalence of 7% for AR among a UAE adult sample across the seven emirates (Mahboub et al. 2014). According to the WHO Allergic Rhinitis and its Impact on Asthma (ARIA) program, patients with persistent and severe rhinitis are more likely to have comorbidity with asthma. Several other epidemiological studies have also suggested a strong association between asthma and rhinitis (Pawankar et al. 2008; Vujnovic and Domuz 2018). Up to 70% of people surveyed with ECRHS who reported asthma also recorded hay fever (Mahboub et al. 2014).

Atopic dermatitis (AD), also known as atopic eczema, is a chronic skin condition that makes the skin red, dry, itchy, and cracked. Globally, around 2–5% of children and 10% of adults suffer from AD (World Allergy Organization 2014). Generally, AD first appears in childhood and flares throughout life (National Eczema Association 2020). The epidemiology of eczema or AD is less recognized, although geographic variations in prevalence are closely related to variations in AR (ISAAC 1993).

### 9.1.2 *Impacts of Respiratory Conditions and Allergies*

Increased incidence of asthma and allergic disorders contributes to increased risk of morbidity and mortality and reduced quality of life (WHO 2019). According to WHO estimates, 417,918 deaths and 24.8 million disability-adjusted life years (DALYs) were attributed to asthma in 2016 (GBD 2016 Disease and Injury Incidence and Prevalence Collaborators 2017; Razzak et al. 2020). Asthma accounts for up to 0.7 deaths in children per 100,000 people and ranks among the top 20 conditions worldwide for DALYs in children of all ages. In older children (10–19 years), asthma has become a more common cause of DALYs over the last decade (Serebrisky and Wiznia 2019).

The financial burden of asthma and other related allergies is relatively high, particularly since asthma often requires hospitalization. Morbidity and comorbidity of respiratory allergies result in frequent emergency visits, exacerbation of symptoms and related hospitalizations, and higher allergy-related medical costs. In general, asthma-related morbidity accounts for 1–2% of healthcare expenditure in developed countries. The direct yearly cost of asthma in Europe reached up to 7.9 billion euros and $37.2 billion in the USA (Mahboub et al. 2013).

The UAE spends substantially more on asthma management and treatment than Europe. For example, the annual cost of asthma in Dubai has been estimated at 88 million dirhams ($24 million), primarily due to outpatient visits and asthma-related hospitalization (Mahboub et al. 2013). The indirect costs include loss of days of work or school and poor quality of life. Asthmatic children have poorer school performance than non-asthmatic children, with nearly half of asthmatic children having learning disabilities (Bener 2011). ECRHS data shows that asthma symptoms led to 52.8% of asthmatic children in the UAE missing school in 2012 and 17.1% of asthmatic adults missing work. Similarly, among asthmatics in Dubai, 14% of adults and 50% of children reported missing a day at work and school, respectively (Mahboub et al. 2013).

### *9.1.3 Risk Factors and Triggers Linked to Respiratory Conditions and Allergies*

Epidemiological research indicates that allergic disorders are much more prevalent in more developed world regions, partly due to modernization and industrialization and adaptation to sedentary lifestyles. Environmental exposure, genetics, and atopy seem to contribute to the globally increasing prevalence of respiratory and allergic diseases (Dharmage et al. 2019; Bonamonte et al. 2019). Historically, the UAE's economy mainly relied on crude oil exports and natural gas production, although more recently it has invested heavily in tourism, manufacturing, and services through massive port facilities and free trade zones. Rapid economic and income growth has resulted in marked changes in lifestyle and the physical and social environment. This in turn has changed the patterns of health and diseases in the different geographical regions of the UAE. Each emirate has a unique environment, demographic composition, and resources, which influence the health of its population. In general, the UAE has a desert climate, with very mild winters and very hot and humid summers with frequent sandstorms (Agency C.I. 2010). These conditions, in combination with exposure to air pollutants from indoor and outdoor sources, may exacerbate chronic respiratory conditions and allergies.

While the specific contribution of genetics to respiratory diseases and allergies is still under investigation, differences in the prevalence of respiratory allergies among various ethnic groups may be due to genetic susceptibility and atopy. Although environmental and genetic vulnerability to allergic diseases exists, these factors do not completely explain major geographic differences in disease prevalence among people with a similar genetic heritage, nor do they account for the global increase in allergic diseases. Other risk factors may include premature birth, age, residential exposure, exposure to diverse chemicals and allergens, and indoor and outdoor air pollutants (WHO 2019).

While there is global data on the prevalence of asthma and other allergic disorders, many countries still lack essential epidemiological data on the specific burden of these diseases. Given the UAE population's heterogeneous demographics and geographic influences, more research needs to focus on the distribution of respiratory diseases and allergies and longitudinal trends across the UAE. In particular, the respiratory health of children and youth remains underexplored. Accordingly, the main aim of this study is to provide an overview on the burden of respiratory conditions and related allergies and to assess any differences in the spatial distribution of these conditions across the UAE. In this chapter, we summarize the results of a study that examined the prevalence and geographic distribution of asthma and other allergies among adolescents residing in the UAE. This chapter has the following main objectives:

I. To examine spatial differences in the prevalence of respiratory conditions and symptoms, symptoms of AR, and symptoms of AD among the UAE adolescent population.
II. To examine differences in the prevalence of respiratory conditions and symptoms, symptoms of AR, and symptoms of AD in the local Emirati versus expatriate subpopulations.

The chapter contributes to the general understanding of the prevalence of asthma and other allergic conditions and risk factors for young adults in the UAE. This is particularly important for developing effective control guidelines to reduce the burden of these disorders and their negative impact on the society and economy.

## 9.2 Method

### *9.2.1 Setting and Study Design*

Respiratory health data were obtained from the National Study of Population Health in the UAE (*NSPHUAE*) research program, conducted between 2007 and 2009. This program was carried out in collaboration with the UAE Ministry of Education (MOE). A total of 147 private and public schools in the 7 emirates, distributed across 9 educational zones, were selected using a stratified sampling strategy based on the UAE MOE's school enrolment data. Schools were randomly chosen by lottery from the 2005–2006 MOE list of schools. A fishbowl lottery method was then used to select one class from each grade (10–12) in the selected schools with more than one class per grade (Barakat-Haddad 2013).

A cross-sectional survey was developed and administered to 6363 adolescents aged 13–20 who attended the selected schools. The survey was divided into two components: the first component was conducted during a 1-h class period by the students and gathered data on lifestyle and physical activity, self-reported medical diagnoses, respiratory health symptoms, injuries, functional capacity and quality of life measures, healthcare access, and demographic and socioeconomic information. The second component of the survey was sent home for parental assistance to collect data related to residential history, characteristics, and environmental exposure (Barakat-Haddad et al. 2015).

### 9.2.2 *Ethical Approval*

Official revision and evaluation of the survey were conducted by the MOE's Office of Research. Approval and ethical clearance were obtained from both the MOE and the Ministry of Health (MOH) prior to the study. To maintain a uniform survey procedure across the emirates, several workshops were conducted to train social workers employed by the MOE on the administration of the survey, emphasizing the importance of accurate and consistent reporting. Detailed letters of information were sent to the adolescent participants' parents or guardians 2 days prior to administering the survey. These letters highlighted the aims and methodology of the research and required the participants' consent (Siddiqua et al. 2018).

## 9.3 Study Instrument

### 9.3.1 *Outcome Variables*

Information on the respiratory health of the UAE adolescents was collected using core questions from the standardized International Study of Asthma and Allergy in Childhood (ISAAC) phase I. The following eight questions from ISAAC questionnaire were used to assess the prevalence of asthma in this population: ever having asthma, ever wheezing, current wheezing or whistling in the chest, current number of wheeze attacks, current sleep disturbance due to wheezing, and current wheezing during or after exercise. The term "current" symptoms refer to symptoms in the past 12 months. To assess the severity of wheezing, adolescent participants were asked if wheezing has ever been severe to limit speech to only one or two words at a time between breaths. Participants were also asked about current nocturnal dry cough apart from cold or chest infection as another presentation of severe asthma. One question addressed hospitalization due to asthma or any other respiratory problem.

Five items assessed the presence of AR or hay fever through responses on whether participants experience certain symptoms when they do not have cold or flu. These symptoms include ever having a problem with sneezing or a runny or blocked nose, current problems with runny or blocked nose, presence of itchy-watery eyes with nose problems, seasonal variation in nose problems, and level of interference of nose problems with daily activities.

Prevalence of atopic eczema and related symptoms was addressed through responses to five items. Participants responded to whether they ever had an itchy rash coming and going for at least 6 months or whether they had an itchy rash at any time in the last 12 months; had an itchy rash that affects the folds of the elbows, behind the knees, in front of the ankles, under the buttocks, or around the neck, ears, or eyes; and a rash clearing completely in the past 12 months. Severity of eczema symptoms was assessed by asking if they had stayed awake at night due to itchy rash in the past 12 months.

#### 9.3.1.1 Independent Variables

Information related to the adolescent participants' demographic profile was based on responses to nine items regarding age, sex, school type, residency, ethnicity (categorized into UAE national versus expatriate from GCC (Gulf Cooperation Council) countries or Greater Arab Free Trade Area, Arab/Middle East, Arab/Africa, Southeast Asia, Western, no nationality, or others). Participants' socioeconomic profile was assessed based on maternal and paternal highest level of education, monthly household income, residential tenure, and residential overcrowding.

### *9.3.2 Statistical Analysis*

The survey data were entered and analyzed using the Statistical Package for the Social Sciences version 20 (SPSS Inc., Chicago IL). Frequency analyses were run for all variables for the entire sample and aggregated by ethnicity (Emirati versus expatriate) and educational zones. Bivariate analyses were performed to identify significant differences between aggregates. Chi-square values were used to measure associations, with p-value <0.05 considered significant.

## 9.4 Results

A total sample of 6363 adolescents aged 13–20 years residing in the UAE were included in this study. The overall response rate was 65%. Detailed information on recruitment and response rate can be found in a previous publication (Barakat-Haddad 2013). Overall, half of the study sample were UAE nationals, and 55% of adolescent participants were females. The mean age was 16 years with 30% of participants being at that age. This suggests that the study sample is representative of the UAE population for that age group. Table 9.1 provides a summary of the main

**Table 1** Sociodemographic and residential profile of adolescent study participants by citizenship status ($n = 6363$)

| Variable | Classification | Overall *n* | Overall % | UAE national (local) *n* | UAE national (local) % | Non-UAE national (expat) *n* | Non-UAE national (expat) % |
|---|---|---|---|---|---|---|---|
| Citizenship status | | | | 3066 | 49.3 | 3157 | 50.7 |
| Sex | Male | 2791 | 44.7 | 1216 | 40.3 | 1515 | 48.3 |
| | Female | 3458 *** | 55.4 | 1798 | 59.7 | 1620 | 51.7 |
| School | Public | 3820 | 61.1 | 2773 | 90.4 | 1043 | 33.0 |
| | Private | 2429 *** | 38.9 | 293 | 9.6 | 2114 | 67.0 |
| Emirate/educational zone | Abu Dhabi/Abu | 1451 | 22.8 | 595 | 19.4 | 835 | 26.4 |
| | Dhabi City | 1175 | 18.5 | 575 | 18.8 | 567 | 18.0 |
| | Abu Dhabi/Al Ain | 377 | 5.9 | 139 | 4.5 | 224 | 7.1 |
| | Abu Dhabi/Western | 264 | 4.1 | 101 | 3.3 | 155 | 5.0 |
| | region | 589 | 9.3 | 280 | 9.1 | 302 | 9.6 |
| | Ajman | 521 | 8.2 | 346 | 11.3 | 164 | 5.2 |
| | Dubai | 671 | 10.5 | 406 | 13.2 | 243 | 7.7 |
| | Fujairah | 1168 | 18.4 | 535 | 17.4 | 605 | 19.2 |
| | Ras Al Khaimah | 147 | 2.3 | 88 | 2.9 | 58 | 1.8 |
| | Sharjah | | | | | | |
| | Um Al Quwain | | | | | | |
| Ethnicity[a] | UAE | 3066 | 48.2 | 3066 | – | – | - |
| | GCC | 359 | 5.6 | – | | 359 | 11.4 |
| | Arab/Middle East | 976 | 15.3 | – | | 976 | 30.9 |
| | Arab/Africa | 620 | 9.7 | – | | 620 | 19.6 |
| | Southeast Asia | 1043 | 16.4 | – | | 1043 | 33.0 |
| | Western | 82 | 1.3 | – | | 82 | 2.6 |
| | None/other | 76 | 1.2 | – | | 76 | 2.4 |
| Father is a high school graduate | Yes | 2908 *** | 57.8 | 841 | 34.9 | 2067 | 79.0 |
| Mother is a high school graduate | Yes | 2478 *** | 49.0 | 665 | 27.3 | 1806 | 68.9 |
| Monthly household income (AED) | <2 K | 161 | 4.2 | 53 | 3.1 | 107 | 4.9 |
| | 2–5 K | 886 | 23.1 | 257 | 15.1 | 618 | 28.2 |
| | 5–8 K | 672 | 17.5 | 236 | 13.8 | 434 | 19.8 |
| | 8–10 K | 469 | 12.2 | 231 | 13.5 | 236 | 10.8 |
| | 10–12 K | 348 | 9.1 | 178 | 10.4 | 169 | 7.7 |
| | 12–15 K | 313 | 8.2 | 158 | 9.3 | 159 | 7.3 |
| | 15–20 K | 285 | 7.4 | 159 | 9.3 | 125 | 5.7 |
| | >20 K | 701 | 18.3 | 434 | 25.4 | 270 | 12.3 |
| Residential property tenure | Owned | 2687 | 51.0 | 2230 | 88.7 | 416 | 15.6 |
| | Rented | 2578 | 49.0 | 283 | 11.3 | 2259 | 84.4 |
| Residential crowding | Yes | 2052 *** | 60.4 | 845 | 50.3 | 1207 | 70.3 |

[a]UAE, local; GCC, Kuwait, Kingdom of Saudi Arabia, Oman, Qatar, Bahrain, and Yemen; Arab/Middle East, Lebanon, Syria, Jordan, Palestine, and Iraq; Arab/Africa, Egypt, Tunisia, Morocco, Algeria, Libya, Sudan, and Somalia; Southeast Asia, India, Pakistan, Bangladesh, Sri Lanka, Philippines, and Indonesia; Western, Europe, USA, Australia, and Canada; no nationality and others, all other nationalities

*$p < 0.05$, **$p < 0.01$, ***$p < 0.001$

sociodemographic and residential characteristics of the adolescent participants. Of the surveyed sample, the majority attended public schools and lived in Abu Dhabi (61.1% and 22.8%, respectively). Of the expatriates, over half were citizens of Middle Eastern Arab countries (30.9%) or Southeast Asian countries (33%). Parental education was higher among non-UAE nationals (68.9% and 79%), while overall monthly household income was higher among UAE nationals, with 25.4% of UAE nationals reporting a monthly household income of more than 20,000 AED compared to only 12.3% of expatriates. 84.4% of expatriate participants lived in a rented house, and 70.3% shared it with more than two people.

### 9.4.1 Prevalence and Spatial Distribution of Respiratory Conditions and Symptoms

Tables 9.2 and 9.3 present the respiratory profile of the participants. Overall, self-reported prevalence of asthma ever was 12%, while 6% reported having spent a night in hospital for asthma or other respiratory illness. The self-reported prevalence of wheezing ever was 15%, while 11.6% experienced wheeze in the past 12 months, 8.4% reported experiencing one to three attacks, and 2.7% reported four or more attacks in the past 12 months. Overall, 2.3% of adolescents reported experiencing sleep disturbances for one or more nights per week due to wheeze, and 2.8% experienced wheeze that limited speech in the past 12 months. 15.7% experienced wheeze during or after exercise, and 33.4% reported experiencing dry cough at night in the past 12 months.

Overall, the self-reported prevalence of asthma ever was significantly higher among Emiratis (14.1%) than expatriates (10.8%, $p < 0.001$); significant differences were found for the residents of Abu Dhabi City (17.3%), Western region (12.7%, $p = 0.007$), Al Ain (11.5%, $p = 0.012$), and Dubai (12.3%, $p = 0.021$). Hospitalization due to asthma or other respiratory illnesses was also significantly higher among Emirati adolescents (7.2%) than expatriates (4.8%, $p < 0.001$), especially in those residing in Abu Dhabi City, Sharjah, Al Ain, and Dubai (6.9%, $p = 0.002$; 6.9%, $p = 0.11$; 6.4%, $p = 0.004$; 5.7%, $p = 0.001$). Conversely, self-reported ever wheeze and wheeze in the past 12 months were more frequent in expatriates (15.9% and 12.4%, respectively) than Emiratis (14.8% and 11.7%, respectively). UAQ participants reported the highest rates of ever wheeze (23.6%) and wheeze in the past 12 months (14.8%), followed by residents of Abu Dhabi City and Sharjah. Expatriates reported significantly higher prevalence of wheeze that limited speech (18.5%) than Emiratis (13.5%, $p = 0.028$).

Significant geographic differences were found for severe wheeze that limited speech ($p = 0.001$) and dry cough at night ($p = 0.001$) in the past 12 months, with UAQ adolescents reporting the highest prevalence in both indicators (33.3% and 48.2%, respectively).

**Table 9.2** Health profile of adolescent study participants in relation to respiratory conditions and symptoms ($n = 6363$)

| Respiratory indicator | | Overall $n = 6363$ | | Local | | Expat | | Test of independence | |
|---|---|---|---|---|---|---|---|---|---|
| | | *n* | % | *n* | % | *n* | % | *p*-value | Test value |
| Spent night at hospital for asthma or other respiratory illness | Yes | 380 | 6.0 | 221 | 7.2 | 152 | 4.8 | **0.000*** ** | 31.910 |
| Asthma ever | Yes | 763 | 12.0 | 416 | 14.1 | 332 | 10.8 | **0.000*** ** | 15.105 |
| Ever wheeze | Yes | 953 | 15.0 | 438 | 14.8 | 486 | 15.9 | – | – |
| Wheeze in past 12 months | Yes | 741 | 11.6 | 341 | 11.7 | 379 | 12.4 | – | – |
| Number of attacks of wheeze in the last 12 months | 1–3 | 532 | 8.4 | 260 | 76.0 | 257 | 74.5 | – | – |
| | 4–12 | 124 | 1.9 | 51 | 14.9 | 68 | 19.7 | | |
| | More than 12 | 51 | 0.8 | 31 | 9.1 | 20 | 5.8 | | |
| Sleep disturbed due to wheeze in the past 12 months | Never or <1 night/week | 788 | 12.4 | 403 | 84.1 | 385 | 85.0 | – | – |
| | 1+ nights/week | 147 | 2.3 | 76 | 15.9 | 68 | 15.0 | | |
| Wheeze severe enough to limit speech in the past 12 months | Yes | 176 | 2.8 | 76 | 13.5 | 94 | 18.5 | **0.028*** | 4.835 |
| Exercise-induced chest wheeze in the past 12 months | Yes | 999 | 15.7 | 481 | 16.6 | 500 | 16.5 | – | – |
| Dry night cough in the past 12 months | Yes | 2125 | 33.4 | 1045 | 35.4 | 1042 | 34.2 | – | – |
| | <1 night/week | 214 | 3.4 | 115 | 21.6 | 94 | 19.6 | | |
| | Never | 658 | 10.3 | 330 | 61.9 | 312 | 65.0 | | |

*$p < 0.05$, **$p < 0.01$, ***$p < 0.001$

### 9.4.2 Prevalence of Allergic Rhinitis or Hay Fever Symptoms

Overall, 33.2% of adolescents reported ever experiencing problems with sneezing, or runny or blocked nose, 26% reported experiencing these symptoms in the past 12 months, and 13.4% reported having nose problem with itchy and watery eyes in the past 12 months (Table 9.4). Nasal problems were most frequently reported in January (9.1%) (Fig. 9.2). Perceived interference of nose problems with daily activities was reported to be mostly minor (16.8%), with 6.8% reporting moderate to major interference.

No significant differences were found between Emirati and expatriate adolescents in self-reported AR/rhino-conjunctivitis symptoms (sneezing, runny or blocked nose symptoms). Significant associations were found between all reported

**Table 9.3** Health profile of adolescent study participants by educational zone in relation to respiratory conditions and symptoms ($n = 6363$)

| Respiratory question | | Total | | Abu Dhabi | | Al Ain | | Western | | Ajman | | Dubai | | Fujairah | | RAK | | Sharjah | | UAQ | | Test of independence | |
|---|---|---|---|---|---|---|---|---|---|---|---|---|---|---|---|---|---|---|---|---|---|---|---|
| | | $N = 6363$ | | $n = 1451$ | | $n = 1175$ | | $n = 377$ | | $n = 264$ | | $n = 589$ | | $n = 521$ | | $n = 671$ | | $n = 1168$ | | $n = 147$ | | | |
| | | *n* | % | *n* | % | *n* | % | *n* | % | *n* | % | *n* | % | *n* | % | *n* | % | *n* | % | *n* | % | *p*-value | Test value |
| Spent night at hospital for asthma or other respiratory illness (%) | Yes | 380 | 6.0 | 98 | 6.9 | 73 | 6.4 | 24 | 6.6 | 10 | 3.8 | 33 | 5.7 | 26 | 5.1 | 26 | 4.0 | 78 | 6.9 | 12 | 8.3 | – | – |
| Asthma ever | Yes | 760 | 12.0 | 243 | 17.3 | 130 | 11.5 | 46 | 12.7 | 30 | 11.9 | 71 | 12.3 | 46 | 9.3 | 46 | 7.1 | 135 | 12.1 | 16 | 11.5 | **0.000** *** | 52.14 |
| Ever wheeze (%) | Yes | 953 | 15 | 242 | 17.2 | 175 | 15.6 | 52 | 14.2 | 37 | 14.7 | 74 | 12.9 | 65 | 13.1 | 81 | 12.7 | 194 | 17.2 | 33 | 23.6 | **0.004** ** | 22.41 |
| Wheeze in the past 12 months (%) | Yes | 741 | 11.6 | 185 | 13.1 | 146 | 12.9 | 38 | 11.2 | 31 | 12.2 | 58 | 10.7 | 47 | 9.3 | 67 | 10.4 | 148 | 13.1 | 21 | 14.8 | – | – |
| # attacks of wheeze in the last 12 months | 1–3 | 532 | 8.4 | 122 | 72.2 | 112 | 80.0 | 26 | 74.3 | 25 | 80.6 | 42 | 75.0 | 39 | 83.0 | 41 | 64.1 | 111 | 76.6 | 14 | 70.0 | – | – |
| | 4–12 | 124 | 1.9 | 34 | 20.1 | 23 | 16.4 | 7 | 20.0 | 3 | 9.7 | 9 | 16.1 | 6 | 12.8 | 14 | 21.9 | 24 | 16.6 | 4 | 20.0 | | |
| | More than 12 | 51 | 0.8 | 12 | 7.7 | 5 | 3.6 | 2 | 5.7 | 3 | 9.7 | 5 | 8.9 | 2 | 4.3 | 9 | 14.1 | 10 | 6.9 | 2 | 10.0 | | |
| Sleep disturbed due to wheeze in the past 12 months | Never or <1 night/week | 811 | 12.5 | 187 | 82.0 | 174 | 86.6 | 46 | 95.8 | 40 | 87.0 | 51 | 79.7 | 61 | 91.0 | 78 | 81.3 | 163 | 86.7 | 11 | 55.0 | 0.002 | 24.92 |
| | 1+ nights/week | 147 | 2.3 | 41 | 18.0 | 27 | 13.4 | 2 | 4.2 | 6 | 13.0 | 13 | 20.3 | 6 | 9.0 | 18 | 18.8 | 25 | 13.3 | 9 | 45.0 | | |

| Respiratory question | | Total N = 6363 | | Abu Dhabi n = 1451 | | Al Ain n = 1175 | | Western n = 377 | | Ajman n = 264 | | Dubai n = 589 | | Fujairah n = 521 | | RAK n = 671 | | Sharjah n = 1168 | | UAQ n = 147 | | Test of independence | |
|---|---|---|---|---|---|---|---|---|---|---|---|---|---|---|---|---|---|---|---|---|---|---|---|
| | | *n* | % | *n* | % | *n* | % | *n* | % | *n* | % | *n* | % | *n* | % | *n* | % | *n* | % | *n* | % | *p*-value | Test value |
| Wheeze severe enough to limit speech in the past 12 months | Yes | 176 | 2.8 | 47 | 18.8 | 28 | 11.4 | 8 | 16.0 | 2 | 3.6 | 16 | 22.9 | 9 | 10.6 | 13 | 12.1 | 45 | 21.1 | 8 | 33.3 | 0.001 | 26.83 |
| Exercise-induced chest wheeze in the past 12 months | Yes | 999 | 15.7 | 241 | 17.3 | 183 | 16.4 | 45 | 12.6 | 39 | 15.9 | 102 | 18.0 | 70 | 14.3 | 88 | 13.9 | 203 | 18.4 | 28 | 20.7 | – | – |
| Dry night cough in the past 12 months | Yes | 2125 | 33.4 | 531 | 38.1 | 391 | 34.7 | 124 | 34.3 | 86 | 34.4 | 192 | 33.4 | 169 | 33.7 | 195 | 30.7 | 370 | 33.0 | 67 | 48.2 | **0.001** ** | 25.44 |
| | <1 night/week | 214 | 3.4 | 40 | 17.9 | 38 | 18.6 | 18 | 30.5 | 11 | 22.0 | 13 | 16.5 | 19 | 20.7 | 23 | 21.7 | 47 | 23.3 | 5 | 22.7 | | |
| | Never | 658 | 10.3 | 153 | 68.3 | 139 | 68.1 | 28 | 47.5 | 32 | 64.0 | 49 | 62.0 | 58 | 63.0 | 75 | 70.8 | 111 | 55.0 | 13 | 59.1 | | |

*RAK* Ras Al Khaimah, *UAQ* Um Al Quwain
$*p < 0.05$, $**p < 0.01$, $***p < 0.001$

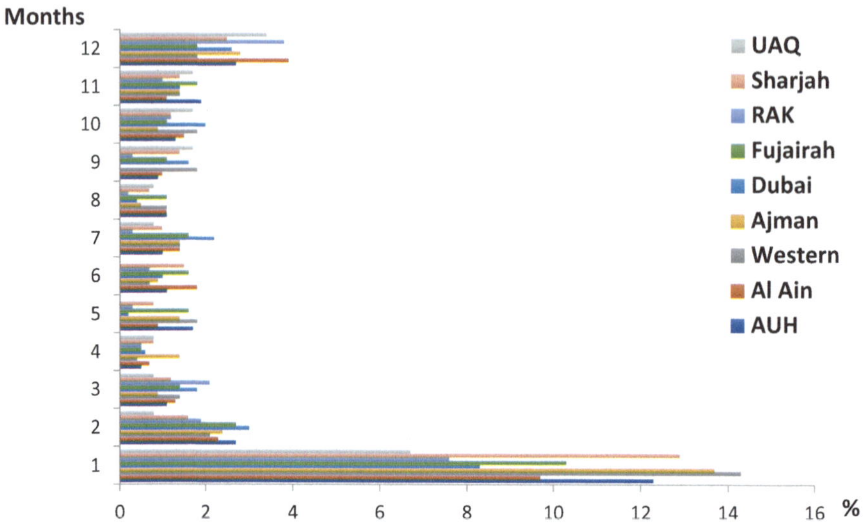


**Fig. 9.2** Temporal profile of nose problem with itchy-watery eyes among UAE adolescents

AR/rhino-conjunctivitis symptoms and emirate of residence (Table 9.5). Specifically, UAQ residents reported significantly higher prevalence than other emirates of ever sneeze or runny or blocked nose (43.3%, $p = 0.001$), mostly among Emiratis ($p = 0.017$). UAQ residents also reported significantly higher prevalence of sneeze or runny or blocked nose in the past 12 months (66.7%, $p < 0.001$), nose problem with itchy-watery eyes (55.4%, $p = 0.003$), and nose problems that moderately to severely interfered with daily activities in the past 12 months (26.5%, $p = 0.009$).

RAK residents reported significantly lower prevalence than other emirates of ever having nose problems with sneezing or runny or blocked nose (26.9%, $p = 0.001$), sneeze or runny or blocked nose in the past 12 months (44%, $p < 0.001$), and nose problem with itchy-watery eyes (28.4%, $p = 0.003$) and nose problems that moderately to severely interfered with daily activities in the past 12 months (16.2%, $p = 0.009$).

### *9.4.3 Prevalence of Atopic Dermatitis/Eczema Symptoms*

Overall, 8.4% of adolescents reported having an itchy rash in the past 6 months, while 7.4% reported experiencing it in the past 12 months, and 5.8% reported that the rash affected the folds of the elbows, behind the knees, ankles, neck, ears, eyes, and buttocks (Table 9.4). Overall, 6.7% of adolescents reported the rash clearing completely in the past 12 months. Of all participants, 3.4% had experienced an itchy rash that kept them awake for one or more nights per week in the past 12 months.

**Table 9.4** Health profile of local and expat adolescents in relation to symptoms of rhinitis or hay fever, atopic dermatitis, and eczema ($n = 6363$)

| Respiratory indicator | | Total ($n$ = 6363) *n* | Total % | Local *n* | Local % | Expat *n* | Expat % | Test of independence *p*-value | Test value |
|---|---|---|---|---|---|---|---|---|---|
| Ever problem with sneezing/runny/blocked nose | Yes | 2112 | 33.2 | 1052 | 35.4 | 1019 | 33.3 | – | – |
| Sneeze/runny/blocked nose in the past 12 months | Yes | 1655 | 26.0 | 808 | 54.4 | 814 | 56.1 | – | – |
| Nose problem with itchy-watery eyes in the past 12 months | Yes | 854 | 13.4 | 424 | 35.2 | 411 | 36.3 | – | – |
| Month-nose problem in the past 12 months | Jan | 580 | 9.1 | 255 | 38.5 | 314 | 43.9 | – | – |
| | Feb | 122 | 1.9 | 75 | 11.3 | 44 | 6.1 | | |
| | Mar | 73 | 1.1 | 35 | 5.3 | 35 | 4.9 | | |
| | Apr | 34 | 0.5 | 16 | 2.4 | 18 | 2.5 | | |
| | May | 56 | 0.9 | 23 | 3.5 | 32 | 4.5 | | |
| | June | 67 | 1.1 | 31 | 4.7 | 34 | 4.7 | | |
| | July | 64 | 1 | 35 | 5.3 | 29 | 4.1 | | |
| | Aug | 45 | 0.7 | 18 | 2.7 | 27 | 3.8 | | |
| | Sep | 57 | 0.9 | 26 | 3.9 | 29 | 4.1 | | |
| | Oct | 74 | 1.2 | 36 | 5.4 | 37 | 5.2 | | |
| | Nov | 79 | 1.2 | 39 | 5.9 | 37 | 5.2 | | |
| | Dec | 153 | 2.4 | 72 | 10.9 | 80 | 11.2 | | |
| Nose problem with daily activities in the past 12 months | Moderate to a lot | 433 | 6.8 | 224 | 19.6 | 209 | 19.9 | – | – |
| | Little | 1067 | 16.8 | 522 | 45.8 | 520 | 49.6 | | |
| | Not at all | 733 | 11.5 | 394 | 34.6 | 320 | 30.5 | | |
| Ever itchy rash in the last 6 months | Yes | 532 | 8.4 | 236 | 8.1 | 289 | 9.5 | **0.047*** | 3.949 |
| Itchy rash in the last 12 months | Yes | 474 | 7.4 | 209 | 24.7 | 257 | 31.2 | **0.003**** | 8.907 |
| Itchy rash affected folds, elbows, or other areas | Yes | 370 | 5.8 | 171 | 26.8 | 194 | 32.6 | **0.025*** | 5.053 |
| Rash cleared completely in the past 12 months | Yes | 426 | 6.7 | 195 | 36.1 | 219 | 44.7 | **0.005**** | 7.872 |
| Stay awake at night due to itchy rash in the past 12 months | 1+ nights/week | 166 | 2.6 | 88 | 16.5 | 74 | 15.4 | – | – |
| | <1 night/week | 214 | 3.4 | 115 | 21.6 | 94 | 19.6 | | |
| | Never | 658 | 10.3 | 330 | 61.9 | 312 | 65.0 | | |

*$p < 0.05$, **$p < 0.01$, ***$p < 0.001$

**Table 9.5** Health profile of adolescents by educational zone in relation to symptoms of rhinitis or hay fever, atopic dermatitis, and eczema ($n = 6363$)

| Respiratory question | | Total | | Abu Dhabi | | Al Ain | | Western | | Ajman | | Dubai | | Fujairah | | RAK | | Sharjah | | UAQ | | Test of independence | |
|---|---|---|---|---|---|---|---|---|---|---|---|---|---|---|---|---|---|---|---|---|---|---|---|
| | | $N = 6363$ | | $n = 1451$ | | $n = 1175$ | | $n = 377$ | | $n = 264$ | | $n = 589$ | | $n = 521$ | | $n = 671$ | | $n = 1168$ | | $n = 147$ | | | |
| | | $n$ | % | $n$ | % | $n$ | % | $n$ | % | $n$ | % | $n$ | % | $n$ | % | $n$ | % | $n$ | % | $n$ | % | $p$-value | Test value |
| Ever problem with sneezing/ runny/ blocked nose | Yes | 2112 | 33.2 | 507 | 36.2 | 393 | 34.9 | 135 | 37.4 | 86 | 34.3 | 190 | 32.5 | 166 | 33.1 | 173 | 26.9 | 400 | 35.5 | 62 | 43.4 | **0.001** ** | 26.48 |
| Sneeze/ runny/ blocked nose in the past 12 months | Yes | 1655 | 26 | 408 | 60.7 | 308 | 51.6 | 109 | 62.3 | 69 | 53.1 | 148 | 60.9 | 121 | 48.2 | 125 | 44.0 | 321 | 55.3 | 46 | 66.7 | **0.000** *** | 41.38 |
| Nose problem with itchy-watery eyes in the past 12 months | Yes | 854 | 13.4 | 207 | 38.1 | 169 | 35.3 | 55 | 39.0 | 39 | 34.8 | 79 | 42.0 | 59 | 29.9 | 60 | 28.4 | 155 | 33.8 | 31 | 55.4 | **0.003** ** | 23.16 |

| Respiratory question | | Total N = 6363 | | Abu Dhabi n = 1451 | | Al Ain n = 1175 | | Western n = 377 | | Ajman n = 264 | | Dubai n = 589 | | Fujairah n = 521 | | RAK n = 671 | | Sharjah n = 1168 | | UAQ n = 147 | | Test of independence | |
|---|---|---|---|---|---|---|---|---|---|---|---|---|---|---|---|---|---|---|---|---|---|---|---|
| | | *n* | % | *n* | % | *n* | % | *n* | % | *n* | % | *n* | % | *n* | % | *n* | % | *n* | % | *n* | % | *p*-value | Test value |
| Month-nose problem in the past 12 months | Jan | 580 | 9.1 | 152 | 43.3 | 95 | 36.0 | 40 | 47.6 | 29 | 49.2 | 42 | 33.3 | 45 | 38.5 | 44 | 37.9 | 125 | 47.2 | 8 | 34.8 | – | – |
| | Feb | 122 | 1.9 | 33 | 9.4 | 23 | 8.7 | 6 | 7.1 | 5 | 8.5 | 15 | 11.9 | 12 | 10.3 | 11 | 9.5 | 16 | 6.0 | 1 | 4.3 | | |
| | Mar | 73 | 1.1 | 14 | 4.0 | 13 | 4.9 | 4 | 4.8 | 2 | 3.4 | 9 | 7.1 | 6 | 5.1 | 12 | 10.3 | 12 | 4.5 | 1 | 4.3 | | |
| | Apr | 34 | 0.5 | 6 | 1.7 | 7 | 2.7 | 1 | 1.2 | 3 | 5.1 | 3 | 2.4 | 2 | 1.7 | 3 | 2.6 | 8 | 3.0 | 1 | 4.3 | | |
| | May | 56 | 0.9 | 21 | 6.0 | 9 | 3.4 | 5 | 6.0 | 3 | 5.1 | 1 | 0.8 | 7 | 6.0 | 2 | 1.7 | 8 | 3.0 | 0 | 0.0 | | |
| | June | 67 | 1.1 | 14 | 4.0 | 18 | 6.8 | 2 | 2.4 | 2 | 3.4 | 5 | 4.0 | 7 | 6.0 | 4 | 3.4 | 15 | 5.7 | 0 | 0.0 | | |
| | July | 64 | 1 | 12 | 3.4 | 14 | 5.3 | 4 | 4.8 | 3 | 5.1 | 11 | 8.7 | 7 | 6.0 | 2 | 1.7 | 10 | 3.8 | 1 | 4.3 | | |
| | Aug | 45 | 0.7 | 14 | 4.0 | 11 | 4.2 | 3 | 3.6 | 1 | 1.7 | 2 | 1.6 | 5 | 4.3 | 1 | 0.9 | 7 | 2.6 | 1 | 4.3 | | |
| | Sep | 57 | 0.9 | 11 | 3.1 | 10 | 3.8 | 5 | 6.0 | 0 | 0.0 | 8 | 6.3 | 5 | 4.3 | 2 | 1.6 | 14 | 5.3 | 2 | 8.7 | | |
| | Oct | 74 | 1.2 | 16 | 4.6 | 15 | 5.7 | 5 | 6.0 | 2 | 3.4 | 10 | 7.9 | 5 | 4.3 | 7 | 6.0 | 12 | 4.5 | 2 | 8.7 | | |
| | Nov | 79 | 1.2 | 24 | 6.8 | 11 | 4.2 | 4 | 4.8 | 3 | 5.1 | 7 | 5.6 | 8 | 6.8 | 6 | 5.2 | 14 | 5.3 | 2 | 8.7 | | |
| | Dec | 153 | 2.4 | 33 | 9.4 | 38 | 14.4 | 5 | 6.0 | 6 | 10.2 | 13 | 10.3 | 8 | 6.8 | 22 | 19.0 | 24 | 9.1 | 4 | 17.4 | | |
| Nose problem with daily activities in the past 12 months | Moderate to a lot | 439 | 6.9 | 96 | 18.6 | 67 | 15.3 | 30 | 22.7 | 27 | 26.7 | 47 | 26.4 | 38 | 20.4 | 33 | 16.2 | 88 | 20.2 | 13 | 26.5 | **0.009**** | 32.36 |
| | Little | 1067 | 16.8 | 266 | 51.6 | 202 | 46.2 | 70 | 53.0 | 43 | 42.6 | 85 | 47.8 | 90 | 48.4 | 94 | 46.1 | 198 | 45.4 | 19 | 38.8 | | |
| | Not at all | 733 | 11.5 | 154 | 29.8 | 168 | 38.4 | 32 | 24.2 | 31 | 30.7 | 46 | 31.2 | 58 | 31.2 | 77 | 37.7 | 150 | 34.4 | 17 | 34.7 | | |
| Ever itchy rash in the last 6 months | Yes | 532 | 8.4 | 133 | 9.6 | 92 | 8.2 | 34 | 9.4 | 16 | 6.6 | 55 | 9.5 | 49 | 10.1 | 47 | 7.4 | 97 | 8.8 | 9 | 6.7 | – | – |

(continued)

**Table 9.5** (continued)

| Respiratory question | | Total $N = 6363$ | | Abu Dhabi $n = 1451$ | | Al Ain $n = 1175$ | | Western $n = 377$ | | Ajman $n = 264$ | | Dubai $n = 589$ | | Fujairah $n = 521$ | | RAK $n = 671$ | | Sharjah $n = 1168$ | | UAQ $n = 147$ | | Test of independence | |
|---|---|---|---|---|---|---|---|---|---|---|---|---|---|---|---|---|---|---|---|---|---|---|---|
| | | *n* | % | *n* | % | *n* | % | *n* | % | *n* | % | *n* | % | *n* | % | *n* | % | *n* | % | *n* | % | *p*-value | Test value |
| Itchy rash in the last 12 months | Yes | 474 | 7.4 | 123 | 32.9 | 73 | 21.7 | 38 | 45.2 | 16 | 18.2 | 49 | 38.9 | 40 | 27.0 | 42 | 23.1 | 83 | 24.9 | 10 | 32.3 | **0.000*** ** | 39.37 |
| Itchy rash affected folds, elbows, or other areas | Yes | 370 | 5.8 | 94 | 34.7 | 56 | 21.6 | 28 | 42.4 | 14 | 21.2 | 40 | 44.0 | 29 | 24.8 | 26 | 21.0 | 72 | 29.5 | 11 | 45.8 | **0.000*** ** | 36.64 |
| Rash cleared completely in the past 12 months | Yes | 426 | 6.7 | 110 | 47.4 | 80 | 36.0 | 26 | 42.6 | 14 | 28.6 | 30 | 40.0 | 40 | 42.6 | 38 | 36.9 | 77 | 37.6 | 11 | 57.9 | – | – |
| Stay awake at night due to itchy rash in the past 12 months | 1 + nights/ week | 166 | 2.6 | 31 | 13.8 | 27 | 13.2 | 13 | 22.0 | 7 | 14.0 | 17 | 21.5 | 15 | 16.3 | 8 | 7.5 | 44 | 21.8 | 4 | 18.2 | 0.049 | 26.00 |
| | <1 night/ week | 214 | 3.4 | 40 | 17.9 | 38 | 18.6 | 18 | 30.5 | 11 | 22.0 | 13 | 16.5 | 19 | 20.7 | 23 | 21.7 | 47 | 23.3 | 5 | 22.7 | | |
| | Never | 658 | 10.3 | 153 | 68.3 | 139 | 68.1 | 28 | 47.5 | 32 | 64.0 | 49 | 62.0 | 58 | 63.0 | 75 | 70.8 | 111 | 55.0 | 13 | 59.1 | | |

*RAK* Ras Al Khaimah, *UAQ* Um Al Quwain
$*p < 0.05$, $**p < 0.01$, $***p < 0.001$

Expatriates reported a significantly higher prevalence of itchy rash in both the past 6 months (9.5%) and 12 months (31.2%) than Emiratis (8.1%, $p = 0.047$; 24.7%, $p = 0.003$, respectively), with the highest prevalence in the Western regions (9.4% and 45.2%). Expatriates reported rashes that affected folds, elbows, and other areas significantly more frequently (32.6%) than their Emirati peers (26.8%, $p = 0.025$), with the highest prevalence in UAQ (45.8%, $p = 0.043$). Significantly, more expatriates (44.7%) reported that rashes cleared completely in the past 12 months than Emiratis (36.1%, $p = 0.005$).

Location of residence was significantly correlated with experiencing itchy rash in the past 12 months ($p < 0.001$). Residents of the Western Emirates and Dubai reported the highest prevalence of itchy rash in the past 12 months (45.2%, 38.9%, respectively), whereas residents of Ajman reported the lowest prevalence (18.2%, $p < 0.001$). 45.8% of UAQ residents who experienced itchy rash reported that it affected folds, elbows, and other areas in the past 12 months ($p < 0.001$). Dubai and Western Emirates residents reported the second and third highest prevalence of itchy rash that affected folds, elbows, and other areas (44% and 42.2%) and also most frequently reported a severe itchy rash that kept them awake for one or more nights per week (21.5% and 22%, respectively).

## 9.5 Discussion and Conclusions

This national UAE-based study was carried out using core questions from ISAAC (International Study of Asthma and Allergies in Childhood). This is one of the few studies to explore the respiratory health of the adolescent population in the UAE. Overall, prevalence rates in adolescents for asthma, allergic rhinitis, and atopic eczema symptoms varied across the UAE's nine educational zones.

The prevalence rate reported here is comparable to previous physician-diagnosed asthma rates of 13% and 13.6% reported, respectively, for 2300 children (6–13 years old) across the UAE in 2000 and 850 school children living in urban and rural areas in the UAE between 1992 and 1993 (al-Maskari et al. 2000). Similarly, previous studies reported high prevalence rates of AR, from 22.9% to 36% in different areas of the UAE (Corbo et al. 1989; Mahboub et al. 2014).

According to ISAAC (2000–2003), conducted in centers worldwide, the prevalence of asthma ever and current asthma varies substantially between centers and countries, with the highest prevalence (≥20%) in Australia, Europe, North America, and parts of Latin America. The current study found more moderate rates of asthma ever, wheeze ever, and current wheeze (12%, 15%, and 11.6%, respectively), which were lower than the global prevalence of current asthma symptoms (14.1%) reported by ISAAC phase III for the 13–14 year age group (Pearce et al. 2007; Mallol et al. 2013). Symptoms of severe asthma, including four or more attacks of wheeze (2.7%), sleep disturbance (2.3%), and limited speech (2.8%), were also lower in the UAE population than the ≥7% prevalence found in many centers worldwide.

Hospitalization due to asthma or other related respiratory symptoms is a key indicator of the burden of severe asthma and improved asthma care systems. Asthma contributes to an estimated 500,000 hospitalization per year, and it is the third-ranking cause of hospitalization among children under the age of 15 (AlBehandy et al. 2015). Results of this study suggest that the prevalence of hospitalization due to asthma and/or other respiratory symptoms was approximately 6%, which may relate to the relatively lower prevalence of severe asthma symptoms in the UAE.

A significantly higher proportion of Emirati adolescents than expatriates were hospitalized due to asthma and other respiratory symptoms (7.2% versus 4.8%) and ever experienced asthma (14.1% versus 10.8%). On the other hand, a significantly higher proportion of expatriate adolescents than Emiratis reported a current severe wheeze that limited speech (18.5% versus 13.5%). These contradictory trends in hospitalization rates and severe wheeze symptom rates may be attributed to differences in access and affordability of the healthcare services among the two populations and across the different emirates. Despite major advances in the UAE's healthcare system, health inequities remain between various income and ethnic groups. After the policy of free healthcare for all residents in the UAE was terminated in 2001, healthcare services continued to be free for Emirati citizens, whereas expatriates were covered by employer-paid health insurance. However, these insurance schemes range widely between basic coverage, mainly for semi-skilled laborers and lower-paid employees, and enhanced coverage, mainly for higher-skilled expatriate workers (Koornneef et al. 2012; Barakat and Siddiqua 2015).

The total cost of asthma in Dubai was estimated at 88 million dirhams ($24 million), with outpatient visits and hospital stays accounting for the largest share of spending, at 37% and 23%, respectively, followed by emergency room (ER) visits and medication costs, at 16% and 20%, respectively. Data from the ECRHS show a high percentage of ER visits (44% of adults and 38% of children) and overnight hospitalizations (28% adults and 38% children) by asthmatics in Dubai, at considerably higher levels than in Europe (Mahboub et al. 2013).

Reported AR symptoms among the UAE's adolescent population reveal high prevalence rates for AR ever (33.2%) and current AR (26%). The prevalence of rhino-conjunctivitis (13.4%) was also higher than the global average (8.5%) for the 13–14 age group reported by ISAAC phase III. Conversely, minor interference of allergic rhinitis symptoms in daily activities (16.8%) was reported by this population. AR symptoms varied seasonally, with peaks in January across all emirates. Eczema prevalence was comparable to the world average of 7.9%, with moderate prevalence (7.4%) for current eczema (Mallol et al. 2013).

We previously found that indoor and outdoor air quality and behavioral practices were significant predictors of respiratory symptoms among this population (Barakat-Haddad et al. 2015; Barakat-Haddad and Zhang 2015). More specifically, data from the NSPHUAE program revealed that several respiratory symptoms are more likely in people living near industrial plants, gas stations, dumpsite, or construction sites; residing near overhead power line/plant; exposed to tobacco and arts, crafts, ceramics, and stains; inhaling fumes of gasoline, glue, correctors/car exhaust, burning ants, and residential exposure. Income was a predictor of spending a night at the hospital for respiratory conditions among Emiratis ($p < 0.05$). In addition, age was found to be a predictor of wheeze among both Emiratis and expatriates, while having asthma ever was a predictor of the intensity of wheeze (number of wheeze attacks) ($p < 0.05$). Residential crowding, experiencing sleep disturbance due to wheeze, and increased number of wheeze attacks were predictors of wheeze that limited speech among expatriates ($p < 0.001$). Barakat-Haddad and Zhang (2015) showed that being Emirati was a predictor of having asthma ever in Dubai and Al Ain; gender predicted asthma in Abu Dhabi; type of school attended was a predictor of asthma in Ras Al Khaimah and Sharjah; being a female Emirati was a predictor of dry nocturnal cough in Al Ain.

In conclusion, this study explored several aspects of the respiratory profiles of UAE adolescents. The prevalence rates of various respiratory symptoms varied across the seven UAE emirates and nine educational zones. Interestingly, adolescents residing in UAQ reported significantly higher prevalence rates of asthma and AR symptoms. UAQ is among the smallest emirate and characterized by its desert land. Over the past two decades, UAQ has witnessed development in the construction sector across the emirate. Exposure to environmental pollution associated with constructions may have contributed to exacerbation of respiratory symptoms among this population.

While the use of the ISAAC-validated questionnaire and the recruitment of a large sample size are strengths of this study, several limitations need to be noted. First, it relied on self-reported symptoms that may be subject to response bias. Second, the sample underrepresented adolescents attending private schools in Dubai and male adolescents in UAQ, which may have influenced the reported prevalence. Third, the survey was distributed with the assistance of social workers in schools, which may have influenced the reporting of some symptoms and behaviors, such as tobacco use. Overall, however, this study provided a detailed understanding of adolescents' current respiratory health in the UAE and demonstrated variations in symptoms across ethnicities and education zones.

## Reflection Questions

1. Discuss the potential risk factors and triggers of respiratory allergies and diseases in the United Arab Emirates.
2. ISAAC is an epidemiological survey used to evaluate the prevalence of atopic diseases in various populations. Discuss the strengths and limitations of ISAAC-derived data for evidence-informed public health measures.
3. Why is it useful to assess geographic variations in health and disease?
4. Using data from developed and developing countries, discuss the economic burden of respiratory condition and allergies.
5. "The COVID-19 pandemic is likely to deepen inequities in respiratory health directly through effects of the disease and indirectly through lockdown measures

exacerbating risk factors and social determinants for respiratory disease" (The Lancet Respiratory Medicine (2020), 8, 8, PP743).

Discuss the effects of COVID-19 pandemic on the respiratory health of disadvantaged groups with reference to risk factors and social determinants of respiratory health.

## References

Agency, C.I. (2010). *The World Factbook: 2010 Edition (CIA's 2009 Edition).*

Al Ghobain, M.O., Al-Hajjaj, M.S. and Al Moamary, M.S. (2012). Asthma prevalence among 16- to 18-year-old adolescents in Saudi Arabia using the ISAAC questionnaire. *BMC Public Health*, 12(239), pp. 1–5. https://bmcpublichealth.biomedcentral.com/articles/10.1186/1471-2458-12-239.

AlBehandy N. S., Hussein H, Al Faisal W., El Sawaf E., Wasfy A., Alshareef N., Altheeb A. and A. S. (2015). Prevalence of Bronchial Asthma and Its Association with Obesity and Overweight Among Adolescents in Dubai, UAE. *International Journal of Preventive Medicine Research*, 1(3), pp. 118–125. http://www.aiscience.org/journal/ijpmr.

al-Maskari, F., Bener, A., al-Kaabi, A., al-Suwaidi, N., Norman, N. and Brebner, J. (2000). Asthma and respiratory symptoms among school children in United Arab Emirates. *Allergie Et Immunologie*, 32(4), pp.159–163. https://pubmed.ncbi.nlm.nih.gov/10900696/.

Alsowaidi, S., Abdulle, A. and Bernsen, R. (2010). Prevalence and Risk Factors of Asthma among Adolescents and Their Parents in Al-Ain (United Arab Emirates). *Respiration*, 79(2), pp.105–111.

Alzaabi, A., Idrees, M., Behbehani, N., Khaitov, M.R., Tunceli, K., Urdaneta, E., Awad, N. and Safwat, M. (2018). Cross-sectional study on Asthma Insights and Management in the Gulf and Russia. *Allergy and Asthma Proceedings*, 39(6), pp.430–436. https://www.ncbi.nlm.nih.gov/pmc/articles/PMC6212633/.

Ann, E., Clin, A. and Arabia, S. (2015). Aeroallergens, atopy and allergic rhinitis in the Middle East. *Eur Ann Allergy Clin Immunol*. 48(1), pp. 5–21. http://www.eurannallergyimm.com/cont/journals-articles/402/volume-aeroallergens-atopy-allergic-rhinitis-middle-1071allasp1.pdf.

Barakat C. and Siddiqua A. (2015). Primary health care use and health care accessibility among adolescents in the United Arab Emirates. *WHO EMRO*, 2(3), pp. 171–184. Available at: http://www.emro.who.int/emhj-volume-21-2015/volume-21-issue-3/primary-health-care-use-and-health-care-accessibility-among-adolescents-in-the-united-arab-emirates.html

Barakat-Haddad, C. (2013). Prevalence of high blood pressure, heart disease, thalassemia, sickle-cell anemia, and iron-deficiency anemia among the UAE adolescent population. *Journal of Environmental and Public Health*, [online] Volume (2013). Available at: https://pubmed.ncbi.nlm.nih.gov/23606864/.

Barakat-Haddad, C. and Zhang, S. (2015). Geographic variations in the predictors of asthma, wheeze, and dry nocturnal cough among adolescents from the United Arab Emirates. *Journal of Public Health and Epidemiology*, 7(4), pp.122–137. https://academicjournals.org/journal/JPHE/article-references/934B21C51268.

Barakat-Haddad, C., Zhang, S., Siddiqua, A. and Dghaim, R. (2015). Air Quality and Respiratory Health among Adolescents from the United Arab Emirates. *Journal of Environmental and Public Health*, [online] Volume (2015). Available at: https://www.hindawi.com/journals/jeph/2015/284595/.

Bener, A. (2011). The Impact of Asthma and Allergic Diseases on Schoolchildren: Are They at Increased Risk of Absenteeism and Poor School Performance? In: A.M. Moldoveanu,ed., *Advanced Topics in Environmental Health and Air Pollution Case Studies*. IntechOpen, number 1113. Available at: https://www.intechopen.com/books/advanced-topics-in-environmental-health-and-air-pollution-case-studies/the-impact-of-asthma-and-allergic-diseases-on-schoolchildren-are-they-at-increased-risk-of-absenteei.

Bener A, Abdulrazzaq YM, Debuse P, al-Mutawwa J. (1994). Prevalence of asthma among Emirates school children. *Eur J Epidemiol*, 10, pp.271–278. Available at: https://pubmed.ncbi.nlm.nih.gov/7859837/

Bonamonte, D., Filoni, A., Vestita, M., Romita, P., Foti, C. and Angelini, G. (2019). The Role of the Environmental Risk Factors in the Pathogenesis and Clinical Outcome of Atopic Dermatitis. *BioMed Research International*, [Online] Volume 2019. https://www.ncbi.nlm.nih.gov/pmc/articles/PMC6500695/pdf/BMRI2019-2450605.pdf.

Corbo, G.M., Ferrante, E., Macciocchi, B., Foresi, A., Angelis, V., Fabrizi, G. and Ciappi, G. (1989). Bronchial hyperresponsiveness in atopic dermatitis. *Allergy*, 44(8), pp.595–598.

Dharmage, S.C., Perret, J.L. and Custovic, A. (2019). Epidemiology of Asthma in Children and Adults. *Frontiers in Pediatrics*, 7(246).

Fadhil, I., Belaila, B. and Razzak, H. (2019). *National accountability and response for noncommunicable diseases in the United Arab Emirates*. 4(1), pp.4–9. Available at:http://www.ijncd.org/text.asp?2019/4/1/4/254492.

GBD 2016 Disease and Injury Incidence and Prevalence Collaborators (2017). Global, regional, and national incidence, prevalence, and years lived with disability for 328 diseases and injuries for 195 countries, 1990–2016: a systematic analysis for the Global Burden of Disease Study 2016. *Lancet (London, England)*, 390(10100), pp.1211–1259. Available at: https://www.ncbi.nlm.nih.gov/pubmed/28919117/.

GMI Blogger (2019). UAE Population Statistics in 2019 (Infographics), GMI. *Official GMI Blog*. Available at: https://www.globalmediainsight.com/blog/uae-population-statistics/.

Hussain, S.M., Farhana, S. A., and Alnasser, S.M. (2018). Time Trends and Regional Variation in Prevalence of Asthma and Associated Factors in Saudi Arabia: A Systematic Review and Meta-Analysis. *BioMed Research International*. [online] Volume 2018. Available at: https://www.hindawi.com/journals/bmri/2018/8102527.

Intranational Study of Asthma and Allergies in Childhood (ISAAC) manual (1993). (2nd edition) Auckland (NZ) / Münster (FRG). Available at: http://isaac.auckland.ac.nz/phases/phaseone/phaseonemanual.pdf.

Koornneef, E.J., Robben, P.B.M., Al Seiari, M.B. and Al Siksek, Z. (2012). Health system reform in the Emirate of Abu Dhabi, United Arab Emirates. *Health Policy*, 108(2–3), pp.115–121.

Mahboub, B.H.S.H., Santhakumar, S., Soriano, J.B. and Pawankar, R. (2010). Asthma insights and reality in the United Arab Emirates. Annals of Thoracic Medicine, 5(4), pp.217–221. https://www.ncbi.nlm.nih.gov/pmc/articles/PMC2954375/.

Mahboub, B.H., Al-Hammadi, S., Rafique, M., Sulaiman, N., Pawankar, R., Al Redha, A.I. and Mehta, A.C. (2012). Population prevalence of asthma and its determinants based on European Community Respiratory Health Survey in the United Arab Emirates. *BMC Pulmonary Medicine*, 12(4).

Mahboub, B., Shandy, F.S.M.A., Safarini, B., AbdulAziz, M., Gamal, M., Parakash, V. (2013). Cost of Asthma in Dubai, United Arab Emirates (UAE). *OMICS Journal of Pulmonary & Respiratory Medicine*. 3(2). Available at: https://www.hilarispublisher.com/open-access/cost-of-asthma-in-dubai-united-arab-emirates-uae-2161-105X.1000146.pdf

Mahboub, B., Al-Hammadi, S., Prakash, V.P., Sulaiman, N., Blaiss, M.S., Redha, A.A. and Vats, D.M. (2014). Prevalence and triggers of allergic rhinitis in the United Arab Emirates. *The World Allergy Organization Journal*, 7(1), pp.19–24. https://www.ncbi.nlm.nih.gov/pmc/articles/PMC4118622/.

Mallol, J., Crane, J., von Mutius, E., Odhiambo, J., Keil, U. and Stewart, A. (2013). The International Study of Asthma and Allergies in Childhood (ISAAC) Phase Three: A global synthesis. *Allergologia et Immunopathologia*, 41(2), pp.73–85. https://www.sciencedirect.com/science/article/pii/S0301054612001097.

National Eczema Association (2020). *Atopic Dermatitis*. [Online] available at: https://nationaleczema.org/eczema/types-of-eczema/atopic-dermatitis/.

Pawankar, R., Baena-Cagnani, C.E., Bousquet, J., Walter Canonica, G., Cruz, A.A., Kaliner, M.A., Lanier, B.Q. and Henley, K. (2008). State of World Allergy Report 2008: Allergy and Chronic Respiratory Diseases. *The World Allergy Organization Journal*, 1(1), pp. S4–S17. Available at: https://www.ncbi.nlm.nih.gov/pmc/articles/PMC3666142/.

Pearce, N., Ait-Khaled, N., Beasley, R., Mallol, J., Keil, U., Mitchell, E. and Robertson, C. (2007). Worldwide trends in the prevalence of asthma symptoms: phase III of the International Study of Asthma and Allergies in Childhood (ISAAC). *Thorax*, 62(9), pp.758–766.

Razzak, H., Harbi, A., Ahli, S. and Shelpai, W. (2020). Chronic obstructive pulmonary disease; prevalence and associated risk factors in the United Arab Emirates. *Hamdan Medical Journal*, 13(1), pp.20–26.

Serebrisky, D. and Wiznia, A. (2019). Pediatric Asthma: A Global Epidemic. *Annals of Global Health*, 85(1):6. Available at: https://www.ncbi.nlm.nih.gov/pmc/articles/PMC7052318/

Siddiqua, A., Dghaim, R. and Barakat Haddad, C. (2018). Sociodemographic predictors of tobacco smoking among expatriate and national adolescents in the United Arab Emirates. *Eastern Mediterranean Health Journal*, 24(03), pp.283–294. https://applications.emro.who.int/EMHJ/v24/03/EMHJ_2018_24_03_283_294.pdf.

Vujnovic, S.D. and Domuz, A. (2018). Epidemiological Aspects of Rhinitis and Asthma: Comorbidity or United Airway Disease. *Asthma Diagnosis and Management - Approach Based on Phenotype and Endotype*. https://cdn.intechopen.com/pdfs/61721.pdf.

WHO. (2007). *Global surveillance, prevention and control of chronic respiratory diseases: a comprehensive approach*. World Health Organization. Available at: https://www.who.int/gard/publications/GARD_Manual/en/.

WHO (2017). *Asthma*. World Health Organization. Available at: Available at: https://www.who.int/news-room/fact-sheets/detail/asthma.

WHO (2018). *World Health Organization - Noncommunicable Diseases (NCD) Country Profiles*, 2018. https://www.who.int/nmh/countries/are_en.pdf.

WHO (2019). *Chronic respiratory diseases*. World Health Organization. Available at: https://www.who.int/health-topics/chronic-respiratory-diseases#tab=tab_1.

World Allergy Organization (2014). Allergic Diseases Resource Center. World Allergy Organization. Available at: https://www.worldallergy.org/education-and-programs/education/allergic-disease-resource-center/

Worldometer (2019). *United Arab Emirates Population (2020)*. Available at: https://www.worldometers.info/world-population/united-arab-emirates-population.

# Chapter 10
# The Epidemiology of Diabetes Mellitus Among Adolescents from the Middle East and North Africa

**Susan Yousufzai and Caroline Barakat**

## Abbreviations

| | |
|---|---|
| DKA | Diabetic keto-acidosis |
| FPG | Fasting plasma glucose |
| GCC | Gulf Cooperation Council |
| IDF | International Diabetes Federation |
| IFG | Impaired fasting glucose |
| NSPHUAE | National Study of Population Health in the United Arab Emirates |
| Pre-DM | Prediabetes mellitus |
| T1D | Type 1 diabetes |
| T2D | Type 2 diabetes |
| UAE | United Arab Emirates |
| WHO | World Health Organization |

## 10.1 Introduction

The growing global burden of diabetes mellitus (commonly known as diabetes) has been illustrated repeatedly over the past two decades as one of the greatest health-care problems and challenges to human health in the twenty-first century (Zimmet et al. 2014). Over the course of the past decade, the estimated number of people (20–79 years) living with diabetes has increased by 62% (Saeedi et al. 2019; IDF

S. Yousufzai (✉) · C. Barakat
Faculty of Health Sciences, Ontario Tech University, Oshawa, ON, Canada
e-mail: susan.yousufzai@ontariotechu.net; Caroline.Barakat@ontariotechu.ca



C. Barakat et al. (eds.), *Adolescent Health in the Middle East and North Africa*,
Global Perspectives on Health Geography,
https://doi.org/10.1007/978-3-030-92107-1_10

2019). In fact, diabetes affects just under half a billion people worldwide (463 million), representing 9.3% of the population. This prevalence has significantly exceeded the predicted estimate for 2025 (300 million adults), (King et al. 1998), and is projected to increase by 25% in 2030 and 51% in 2045 (Saeedi et al. 2019). The changing behavioral dynamics of our society has played a contributory etiological role in the expression of this chronic disease, interactively with genetic and environmental risk factors, in both developed and developing countries (Saeedi et al. 2019; Zimmet et al. 2014). Particularly concerning is the increasing prevalence of juvenile diabetes (i.e., type 1 diabetes; T1D) and early-onset of type 2 diabetes (T2D), which has steadily contributed to this burgeoning epidemic across adolescents (10–19 years) (Guariguata and Jeyaseelan 2019).

Diabetes is a complex chronic disease influenced by a multitude of modifiable and nonmodifiable risk factors that can translate into a range of metabolic disorders characterized by high blood glucose levels affecting individuals as young as 6 months of age to individuals 65 years and older (Saraswathi et al. 2019; WHO 2016). T1D has been considered as one of the most common noncommunicable diseases diagnosed during the stages of childhood and adolescence (<19 years of age), with a peak incidence around puberty, and accounts for 5–10% of diabetes cases worldwide (Robert et al. 2018; Tuomilehto 2013). However, T2D, once considered an adult-onset disease, is also prevailing in many areas of the world among adolescents (Farsani et al. 2013), resulting in a shift in the epidemiological patterns of this disease once separated by its historical etiological tendencies. For example, recent national epidemiological studies suggest that the mean age of T2D diagnosis among adolescents is 13.7 years (Panagiotopoulos et al. 2018). According to the Global Burden Report (2019), diabetes affects more than 8.8. million children and adolescents (<20 years) worldwide. Recent prevalence estimates of T1D among adolescents from the International Diabetes Federation (2019) show that the worldwide annual incidence estimates were 98,200 and 128,900 new cases in the under 15-year and under 20-year age-groups, respectively (Patterson et al. 2019). Globally, 1.1 million children and adolescents are affected by T1D (IDF 2019). On the other hand, due to the recent recognition of T2D among adolescents, there is substantial variances and limited data available worldwide (Abuyassin and Laher 2016; Farsani et al. 2013).

Moreover, while the precise etiology and risk factors associated with T1D are not substantially characterized, nor presentably known, the progression from normality to T2D is more preventable. Specifically, T1D is a genetically induced auto-immune condition resulting in the slow destruction of insulin-secreting β cells that regulate glucose in the blood (Rasoul et al. 2016), while T2D is characterized by a progressive decline in insulin production and subsequent inefficiency of the body to utilize produced insulin due to developed resistance. T2D is considered a heterogenous disease, predominantly attributed to recent changes in behavioral and environmental conditions (e.g., urbanization and obesity) (Saeedi et al. 2019). Major risk factors such as reduced physical activity, poor nutritional practices, and in consequence, obesity, have been linked to T2D development in adolescents, as well as early expression in genetically susceptible individuals (Panagiotopoulos et al. 2018).

Certain populations, such as the Arab population, have been considered high risk for T2D. This may increase the risk of early-onset diabetes, given that T2D development is often correlated with a family history of T2D in a first-or second-degree relative (Panagiotopoulos et al. 2018; Zabetian et al. 2013).

### *10.1.1 A Regional Overview of the MENA*

The Middle East and North Africa (MENA) region, widely considered as the "Arab World" has witnessed an increase in diabetes prevalence, aligned with the rapid growth of the region both economically and demographically (Al Makadma 2017; Zabetian et al. 2013). Specifically, the public health concern associated with the rising diabetes prevalence in the MENA region was brought to attention following the appearance of related risk factors associated with the demographic and geopolitical establishment of the region, which translated into changes in societal behaviors in lifestyle (e.g., nutritional practices and physical inactivity), and more positive associations, such as the decrease in perinatal mortality (Zabetian et al. 2013). According to the International Diabetes Federation (2019), the MENA currently has the highest world-age standardized diabetes prevalence (ages 20–79 years), accounting for 12.2% (CI 8.3–16.1) of the population (54.8 million people; CI 30.7–75.1) (Saeedi et al. 2019), a number that is quickly approaching the estimated prevalence expected for 2030 (59.9 million) (Whiting et al. 2011). Correspondingly, this projection may be accelerated due to the increasing prevalence of adolescent diabetes, and further exceed the prevalence in other regions worldwide.

In fact, high-income countries of the MENA region considered a part of the Gulf Cooperation Countries (GCC), such as, Kuwait have witnessed a prevalence much higher than that of Canada and the United States, of 34.9 per 100,000 among adolescents 6–18 years old for T2D (Moussa et al. 2008). Comparatively, in Canada, the national surveillance study demonstrated a minimum incidence of 1.54 per 100,000 in children and adolescents <18 years of age of T2D, while the states demonstrated an incidence of 8.1 per 100,000 person-years in the 10- to 14-year age group and 11.8 per 100,000 person-years in the 15- to 19-year age group (Panagiotopoulos et al. 2018). Moreover, the MENA region holds countries that are ranked among the top 10 for incidence rates of T1D in the 0- to 14-year age group. This includes Kuwait, Saudi Arabia, and Qatar, which also represent the highest prevalence for diabetes cases for individuals 0–19 years in the MENA region (Fig.10.1). Overall, the number of newly diagnosed cases for children and adolescents with T1D each year in the MENA region is 20,800, with currently 149,400 diagnosed (0–19 years) (IDF 2019). Correspondingly, the MENA is also represented as having the third largest estimated deaths in those with T1D (Patterson et al. 2019), indicating a dire neglect associated with diabetes management. This may be a consequence of limited health resources in some developing countries of the MENA such as Jordan, resulting in deterioration of metabolic control (Alassaf et al. 2019). In fact, acute complications, such as diabetic ketoacidosis (DKA)

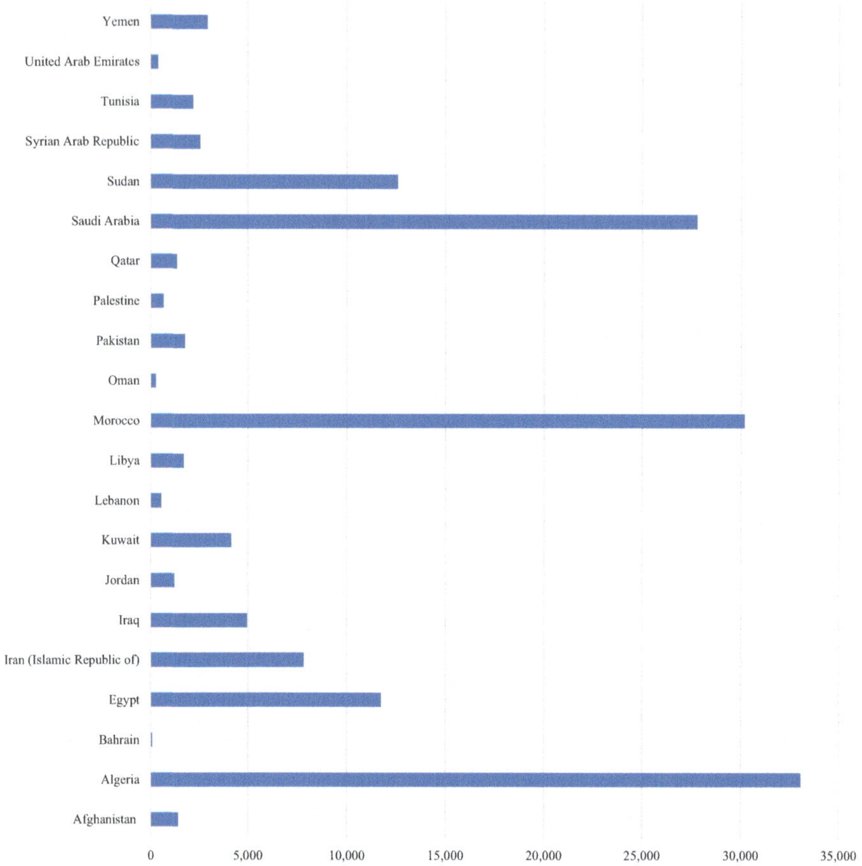


**Fig. 10.1** Country-wise statistics for the prevalence of T1D in the Middle East and North Africa Region, represented as members of the International Diabetes Federation from IDF Diabetes Atlas: MENA 9th edition 2019

associated with T1D among adolescents due to lack of parental awareness, and delayed medical seeking, are common throughout the MENA, and may increase the risk of mortality (Cherian et al. 2010; Robert et al. 2018; Zayed 2016).

### *10.1.2 Living with Diabetes in Adolescence*

Adolescence is recognized as a transitional period of life involving multifactorial changes that are controlled by biological and environmental (i.e., the social, cultural, and built environment) factors (Al Makadma 2017). These excessive and fundamental changes make this cohort vulnerable to the determinants of their social and physical environment, and act as predictors of positive development. Within this transitional phase, adolescents also experience the need for social success through acceptance and require a supportive family and social network to promote effective growth and health (Barakat and Yousufzai 2020). The success of these interactions and fulfilment of this desire may be dependent upon the health-related status of individuals.

Health conditions may negatively impact aspects of this period consisting of physical, emotional, and mental changes which coincide with hormonal shifts during puberty. Specifically, diabetes can have disrupting effects on young individuals during their most productive years, as their time becomes replaced by a rigorous regimen, and the possible emotional burden of coping with discrimination and stigmatization, and increase risk of decline in self-care (Elissa et al. 2017). Certain social and cultural contexts may also extenuate the burden and management of this condition for adolescents and lead to poor health outcomes, especially for those from the Arab culture.

Although feelings of alienation and stigmatization are common among adolescents living with diabetes from Western countries (e.g., the United States) (Buchbinder et al. 2005; Elissa et al. 2017), the experience of adolescents living with diabetes in the MENA region may be further magnified by traditional gender roles, societal norms, and misconceptions. For example, a study conducted in the West Bank of Palestine found that parent's concerns about their female children stemmed from lack of acceptance as a "suitable" wife and perceived inability to uphold traditional roles of Arabic women (e.g., taking care of children and household duties) as a result of their condition. In fact, lack of knowledge and misunderstandings about T1D in society have led to negative consequences like poorer adherence to management of diabetes, especially for females from trying to hide the condition from peers (Elissa et al. 2017). In addition, in some more religiously conservative areas throughout the MENA, social constraints related to physical activity may further restrict females from being able to control their diabetes and lead to increased health complications (Elissa et al. 2017). This may further facilitate differences in gender health outcomes and signifies the need for monitoring of this disease among vulnerable populations, as well as warranting the widespread knowledge of this disease for both parents and adolescents living in the MENA region, in order to mitigate misconceptions influenced by cultural perspectives.

Accordingly, due to the complexity of diabetes and its increasingly prevalent forms (i.e., T1D and T2D) among adolescents, and the lack of understanding on the etiological expression of this condition, more research is required to identify various risk factors that may contribute differently to the geographic variation in prevalence. While literature suggests a range of risk factors that may be correlated to the incidence of diabetes in adolescents, there are limited reviews examining the variation of these factors and related studies in relation to all countries of the MENA region (Saraswathi et al. 2019). Overview of trends in the incidence and prevalence of diabetes forms that may peak in adolescents in the MENA region is essential, given its large youth population and the region's dependence on this "youth bulge" for future development and prosperity. Furthermore, the increasing prevalence of associated modifiable risk factors for early-onset diabetes reinforces the need to create awareness about the formation of poor habits during this critical period.

## 10.2 Methods

This chapter focuses on the prevalence and predictors of diabetes by synthesizing evidence on literature published to date on adolescents (10–19 years) in the MENA region. The following countries identified as members of the International Diabetes Federation within the MENA region were inputted in the search: Afghanistan, Algeria, Bahrain, Egypt, Iran, Iraq, Jordan, Kuwait, Lebanon, Libya, Morocco, Oman, Qatar, Sudan, Saudi Arabia, Syria, Tunisia, the United Arab Emirates, Pakistan, Palestine, and Yemen. A comprehensive search strategy was applied to search for all publications in five databases (OVID Medline, CINAHL, Google Scholar, Web of Science, and Scopus), from the year 2000 to 2020 of each database up to September 2020 limited to the English language. Boolean/phrase search was conducted using the following relevant search terms: "diabetes mellitus" OR "impaired glucose tolerance" OR "type 1 diabetes" OR "type 2 diabetes" AND Adolescen* OR Child* OR Youth AND "prevalence" OR "incidence" AND <country name> OR <region/continent> (e.g., "MENA" OR "Middle East" OR "North Africa" OR "GCC").

The search of the five electronic databases in total yielded 572 results. Article titles and abstracts were downloaded in EndNote. A two-phase screening methodology to identify relevant articles was employed. In phase 1, the reference list was screened for duplicates and each article's title and abstract for eligibility using the criteria described above was screened, resulting in 72 articles. Articles that were possibly relevant (not enough detail was provided to determine eligibility in the title and abstract) were further reviewed in phase 2. In phase 2, the full text of articles was reviewed to determine eligibility. Additional articles were identified based on references from other sources. Identified articles were iteratively screened for relevance regarding adolescents in relation to prevalence, incidence, predictors, and risk factors of diabetes among this population in the MENA region by article title, abstract, and full review of article.

In this chapter, adolescents are defined as the age group from 10 to 19 years old, in accordance with definitions from WHO (2020). However, multiple studies in this review include a wide age spectrum and population samples coinciding with the period of adolescence, which encompassed individuals from 0 to 19 years old (Table 1). Exclusion was based on studies that reported only on pediatric/child populations (<10 years of age), adult populations (>19 years), articles that were not written in English, and review articles. Although only one systematic review was found that specifically focused on childhood (ages 0 and 18 years) diabetes in the Middle East (Saraswathi et al. 2019). In total, 21 studies were eligible for inclusion.

**Table 10.1** Summary of studies in relation to prevalence and incidence rate (IR) of type 1 and type 2 diabetes and other diabetic conditions among adolescents in the MENA region

| | Author/year | Study design | Study population | Region | Diagnosis | Key findings |
|---|---|---|---|---|---|---|
| 1 | Aminzadeh et al. (2019) | Retrospective (2000–2015) | $N = 988$<br>Age:<br><15 years | Iran (Southwest) | T1D | Mean annual IR raised from 18 to 27 to 138 per year during 15 years<br>Annual IR: 13.35/100,000 |
| 2 | Moadab et al. (2010) | Cross-sectional | $N = 672$<br>Age:<br>6–19 years | Iran, Isfahan | IFG and T2D | Prevalence of IFG: 4.61%<br>The prevalence of T2D was 0.1% ($n = 1$; age, 18.00 years old). |
| 3 | Al-Ghamdi and Fureeh (2018) | Retrospective | $N = 471$<br>Age:<br>0–19 years | Saudi Arabia, Al-Baha | T1D | Prevalence: 355/100,000<br>Prevalence was 57.5% vs. 42.5% ($p = 0.3$) for females and males, respectively, and the female/male ratio was 1.36 in favor of females |
| 4 | Al-Daghri et al. (2015) | Cross-sectional | $N = 2225$<br>Age:13–17 years | Saudi Arabia, Riyadh | T2D and prediabetes | Prevalence: 4.3% and 1.6% for males and females, respectively<br>The prevalence of elevated fasting blood glucose levels was 20.5% for females and 17.7% for males ($p < 0.001$) |
| 5 | Al-Rubeaan (2015) | Observational cross-sectional | $N = 23$<br>Age:<br>≤18 years | Saudi Arabia | T1D, T2D, and IFG | Age-adjusted prevalence of T2D: 1/1000<br>Total prevalence: 105 participants (0.45%), of which 88 (0.38%) were T1D and 17 (0.07%) were T2D<br>Of 17,207 (73.48%) participants aged 6–18 years who were tested for FPG, 1053 (6.12%) were IFG cases, and 735 (4.27%) were newly identified cases with diabetes totaling 10.39% |
| 6 | Al-Agha et al. (2012) | Retrospective cross-sectional study | $N = 387$<br>Age:<br>2–18 years | Saudi Arabia | Hyperinsulinism and T2D | Prevalence of hyperinsulinism: 44.7%<br>Prevalence of T2D: 9.04% |

(continued)

**Table 10.1** (continued)

| | Author/year | Study design | Study population | Region | Diagnosis | Key findings |
|---|---|---|---|---|---|---|
| 7 | Habeb et al. (2011) | Observational | $N = 419$<br>Age:<br>0–12 years | Saudi Arabia, Al Madinah | T1D | The mean annual age-standardized IR was 29.0 per 100,000 (95% CI: 26.0–32.0)<br>249 (59.4%) were females and 170 (40.6%) were males, giving a male to female ratio of 1.5:1 |
| 8 | Abduljabbar et al. (2010) | Subpopulation-based observational incidence study (1990–2007) | $N = 438$<br>Age:<br><15 years | Saudi Arabia | T1D | The average IR was 27.52/100,000/year (95% CI: 26.72–28.32), increasing from 18.05/100,000/year in the first 9 years of the study period to 36.99/100,000/year in the last 9 years. |
| 9 | Al-Herbish et al. (2008) | Cross-sectional, case (2001–2007) | $N = 45{,}682$<br>Age:<br>5–18 years | Saudi Arabia | T1D | Fifty were identified as having T1D<br>Prevalence rate: 109.5 per 100,000<br>Prevalence range by region: 48 (eastern) and 162 (central)/100,000<br>Children and adolescents were also grouped by age and prevalence rate: 5–6 (100), 7–12 (109), 13–16 (243), and 17–18 (150) |
| 10 | Shaltout et al. (2017) | Observational, case | $N = 515$<br>Age:<br>0–14 years | Kuwait | T1D | Crude IR: 40.9 (95% CI 37.4–44.6).<br>Age standardized rate: 41.7 (95% CI: 38.1–45.4) per 100,000 per year<br>IR males: 39.2 (95% CI: 34.6–44.4)<br>IR females: 44.1 (95% CI: 39.0–49.7) |
| 11 | Moussa et al. (2008) | Cross-sectional, cohort | $N = 128, 918$<br>Age:<br>6–18 years | Kuwait | T2D | Prevalence of 34.9 per 100,000 [95% CI:24.7–45.1] |
| 12 | Shaltout et al. (2002) | Retrospective, case | $N = 364$<br>Age:<br>0–14 years | Kuwait | T1D | IR: 20.1 per 100,000 (95% CI: 18.0–22.1)<br>Age-standardized rate: 20.9 (95% CI: 18.8–23.0)<br>IR males: 21.1 per 100,000 (95% CI: 18.1–24.1)<br>IR females: 19.0 per 100,000 (95% CI: 16.1–21.8) |

| | Author/year | Study design | Study population | Region | Diagnosis | Key findings |
|---|---|---|---|---|---|---|
| 13 | Barakat et al. (2021) | Cross-sectional | $N = 6329$ Age: 15–18 years | UAE | Diabetes (unspecified) | Prevalence: 0.9% (95% CI: 0.7–1.2)<br>1.5% (95% CI: 1.0–2.1) of males compared to 0.5% (95% CI: 0.3–0.8) of females reported a self-diagnosis of diabetes ($p < 0.001$) |
| 14 | Al Amiri et al. (2015) | Observational/ cross-sectional | $N = 1037$ Age: 11–17 years | UAE, Sharjah | Prediabetes and T2D | 443 (43%) students had abnormal screening results. The prevalence of prediabetes and T2D was 5.4% and 0.87%, respectively, based on a standard oral glucose tolerance test |
| 15 | Punnose et al. (2005) | Case series/ retrospective | $N = 96$ Age: 0–18 years | UAE, Al-Ain | T2D | 11 were identified as type 2 |
| 16 | Punnose et al. (2002) | Retrospective analysis (1990–1998) | $N = 40$ Age: 0–18 years | UAE, Al-Ain | T1D and T2D | 35 patients had T1D and five patients had features of early onset T2D |
| 17 | Alyafei et al. (2018) | Prospective cohort study | $N = 485$ Age: $0.5 \leq \text{age} \leq 14$ | Qatar | T1D and T2D | T1D in this population over the period 2012–2016 was 28.39/100,000 (95% CI: 31.82–40.03). This was significantly higher compared to the unadjusted estimated incidence registered between 2006 and 2011 (23.15/100,000)<br>The incidence of T2D increased from 1.82 per 100,000 in 2012 to 2.7 per 100,000 in 2016, with an incidence of T2D equal to 2.9/100,000 per year |
| 18 | Ali et al. (2013) | Comparative cohort study | $N = 210$ Age: 1–18 years | Egypt | T2D | Frequency: 28 (13.3%) |
| 19 | Osman et al. (2013) | Retrospective, descriptive hospital study (2006–2009) | $N = 958$ Age: 11–8 years | Sudan, Khartoum | T2D | Frequency: 38 (4%) |

## 10.3 Results

The results section includes the prevalence and incidence rates for T1D and T2D from countries identified in Table 1. The most predominant risk factors and characteristics identified in studies for T1D and T2D development are also discussed, which include, pubertal onset, vitamin D deficiency, sex, obesity, and family history of diabetes.

### *10.3.1 Prevalence and Incidence Rates*

#### T1D

Studies specifically reporting on the incidence and prevalence rates of T1D were conducted in Saudi Arabia (5), Kuwait (2), the United Arab Emirates (UAE) (2), Iran, and Qatar.

***Saudi Arabia*** The incidence and prevalence of T1D was illustrated in five studies from Saudi Arabia and it shows an increasing trend over the years in prevalence and incidence rates for adolescents with T1D. Two observational case studies with a large study range period from 1990 to 2007 and 2004 to 2009 were conducted in Eastern Saudi Arabia and North-west Saudi Arabia, respectively. In the study conducted over a period of 18 years (1990 and 2007), a noticeable increase was found for incidence rate at an average increase of 27.52/100,000/year. The incidence rate was 36.99 per year compared to 18.05, in the first 9 years of study (100,000/year) (Abduljabbar et al. 2010). The second study, conducted over 5 years in Al-Madinah on children below 12 years of age, found a mean annual age-standardized incidence rate of 29 per 100, 000 (95% CI 26.0–32.0), although a significant annual increase in incidence rates was not found (Habeb et al. 2011). Prevalence rates were much higher for two other studies. A larger nationwide Saudi Arabian project conducted between 2001 and 2007 found that the overall prevalence rate was 109.5/100,000 and ranged between 48 and 162 (per 100,000) for individuals 5–18 years (Al-Herbish et al. 2008). The highest prevalence rate was reported for the most recent study conducted in the Al-Baha region in Saudi Arabia. This study found that over 10 years (2007–2016), the prevalence rate of T1D was 355/100,000 population of adolescents (0–19 years) (Al-Ghamdi and Fureeh 2018).

A large cross-sectional study ($n$ = 23,532) reported on the prevalence of various diabetes diagnosis among adolescents (≤18 years), including T1D, T2D, and impaired fasting glucose (IFG). The age-adjusted prevalence of T2D in this study was 1/1000. Specifically, out of 105 adolescents diagnosed with diabetes, 0.07% were T2D and 0.38% were T1D (Al-Rubeaan 2015). However, in this study, more than 90% of the children and adolescents with diabetes were unaware of their disease, of which the known patients with T1D and T2D accounted for only 0.45%, while the newly identified cases with diabetes and IFG accounted for 10.39% (Al-Rubeaan 2015).

***Kuwait*** In Kuwait, an increase in the incidence of T1D among children (<15 years) has been reported when comparing reports from 1980–1981 to 1992–1993, in which the incidence rate grew four times higher, from 3.96/100,000 person-years to 15.4/100,000 (Taha et al. 1983; Shaltout et al. 1995). Two studies reporting incident rates for Kuwait included in this review were conducted between 1992–1997 (Shaltout et al. 2002) and 2011–2013 (Shaltout et al. 2017). These studies reported data on newly diagnosed patients that were registered through the Childhood-Onset Diabetes e-Registry. The most recent study found that the incidence rate found in individuals (aged 0–14 years) was 40.9/100,000 per year from 2011 to 2013, which was 2.3 times higher (95% CI 1.9–2.7) than the incidence rate reported during 1992–1997 (17.7).

***Iran*** A recent study from Iran also found that the annual incidence of T1D for children under 15 years of age, between the years 2000 and 2015, was 13.35/100,000 (Aminzadeh et al. 2019).

***UAE*** One of the first studies to report on the patterns of diabetes incidence in adolescents from the UAE found that over 9 years (1990–1998), 40 adolescents (0–18 years) were diagnosed (Punnose et al. 2002). In this cohort, 35 individuals were diagnosed with T1D. A recent secondary analysis of data from the National Study of Population Health in the UAE, conducted between 2007 and 2009 (NSPHUAE 07–09), found that the prevalence of diabetes (not specified) based on self-reported diagnosis on a randomly selected stratified sample of 6329 school adolescents (15–18 years old) from 9 regions of the UAE was 0.9% (95% CI 0.7–1.2) (Barakat et al. 2021).

***Qatar*** In a study conducted in Qatar, 440 youth with T1D (0.5–14 years) were identified from the registry of the National Pediatric Diabetes Centre. This study was a prospective cohort study which estimated the incidence and trend of T1D during the period between 2012 and 2016 and compared it to the incidence previously reported from 2006 to 2011. Results indicated that the inclusive unadjusted estimated incidence rates in this population was 28.39/100,000, which was significantly higher compared to the estimated incidence registered between 2006 and 2011 (23.15/100,000) (Alyafei et al. 2018). Equally, this study reported on the incidence of T2D, which also increased between 2012 and 2016 from 1.82 to 2.7 per 100,000.

## T2D

The study from Qatar found that the overall incidence of T2D equated to 2.9/100,000 per year (Alyafei et al. 2018). The prevalence and characteristics of adolescents with T2D was also reported in Egypt (Ali et al. 2013), Sudan (Osman et al. 2013),

Kuwait (Moussa et al. 2008), the UAE (Punnose et al. 2005), and Iran (Moadab et al. 2010). In a large cross-sectional study conducted in Kuwait, including 128,918 adolescents (6–18 years old) from 182 schools, the overall prevalence was 34.9/100, 000 (Moussa et al. 2008). A retrospective study from the Al-Ain hospital from the UAE characterized the features of T2D among 96 newly diagnosed children with diabetes, of which 11 were reported to have T2D (Punnose et al. 2005). In Egypt, a comparative cohort study found the prevalence of T2D to be 13.3% among adolescents diagnosed with diabetes (210) (Ali et al. 2013). In Sudan, the prevalence was lower, at 4%, based on the clinical records of 985 children attending a diabetic clinic for children and adolescents between 2006 and 2009. The lowest prevalence of T2D was reported by Moadab et al. (2010) in Iran, where the prevalence of T2D was 0.1% in a large cross-sectional study.

### *10.3.2 Risk Factors and Characteristics for T1D and T2D Among Adolescents*

**Pubertal Onset**

The stage of adolescence is also known to coincide with the presentation of prediabetes symptoms, specifically, insulin resistance which is characterized by a lack of glucose uptake. Temporary insulin resistance is suggested to occur as a result of secretion of specific hormones (e.g., growth hormone) associated with development during puberty (Habeb et al. 2011). While decreased insulin sensitivity is normally compensated by an increase in insulin secretion, this period can make adolescents vulnerable to T2D development, which may be extenuated by the adverse effects of obesity (Kelly et al. 2011). Interestingly, one study from the UAE indicated that adolescents required an increase in insulin at the onset of puberty, and a peak incidence of T1D was in the 10–14 years age group (Punnose et al. 2002). These findings suggest that a critical factor to examine in relation to the incidence of T1D and T2D is the age of onset, and that adolescence may be a particularly vulnerable stage. Indeed, studies report a peak incidence of T2D representation during mid-puberty among adolescents (Arslanian 2000; Reinehr 2013). Similarly, many epidemiological studies have witnessed the disease onset of T1D to occur after the age of 15 years, coinciding with the years of adolescence rather than childhood (Tuomilehto 2013).

Pubertal onset was identified in a number of studies in relation to diagnosis of T2D. Two studies in particular indicated the association between T2D and pubertal onset (Punnose et al. 2005; Osman et al. 2013). In Sudanese adolescents, 92.1% had onset between 11 and 18 years of age and were all considered "pubertal" (Osman et al. 2013). In the study from the UAE, out of the identified adolescents with T2D, the mean age of diagnosis was 14.6 (SD 3.0) years, and 10/11 adolescents were considered to be portraying signs of puberty (Punnose et al. 2005). Similarly, a retrospective study in Saudi Arabia among adolescents with T2D indicated that the mean (SD) age of diagnosis was 13.1 ± 2.02, and 60% were considered pubertal

(Al-Agha et al. 2012). A study conducted in Kuwait indicated that a key pattern of prevalence of T2D in this study was the high percentage of children in the age group 14–18 years (57.8%), as the prevalence rate in this group was the highest (46.8 per 100,000) (Moussa et al. 2008). Conversely, two studies indicated that the highest T1D prevalence was for adolescents in the higher age groups; 13–18 years old (Al-Herbish et al. 2008), and 10–12 years old (Habeb et al. 2011) (Table 2).

## Vitamin D Deficiency

Recent increase in vitamin D deficiency has been linked to a potential influence on several chronic diseases (Al-Daghri et al. 2015). The incidence and association with vitamin D deficiency and diabetes in particular has been a risk factor found among children and adolescents diagnosed with diabetes in Middle Eastern countries, despite the abundancy of seasonal exposure to sunlight (Al-Daghri et al. 2015; Robert et al. 2018). Vitamin D3 supplementation during the intrauterine period and early life may have a protective effect against autoimmune diseases (Cherian et al. 2010). For example, lack of sunlight in countries outside the MENA region, such as Finland, has suggested the relationship between lack of endogenous vitamin D production and exposure in pregnancy to the incidence of diabetes in offspring (Chakhtoura and Azar 2013). In one study from this review, in a sample of 2225 adolescents (13–17 years) from Riyadh with T2D, the difference between females and males with vitamin D deficiency (<25 nmol/L) was statistically significant (females: 47%; males:19.4%). However, vitamin D deficiency was significantly associated with T2D and prediabetes only in males. This indicates a sex-related disadvantage for males with low vitamin D status, and possible etiological differences associated with the incidence of T2D based on sex- and gender-related risk factors (Al-Daghri et al. 2015).

## Sex-Related Prevalence

Studies show variability in the prevalence of diabetes diagnosis in females and males. The prevalence of diabetes diagnosis, specifically T2D, was significantly higher among males (47.3, 95% CI 28.7–65.8) than females (26.3, 95% CI 14.8–37.8) from a study in Kuwait of a randomly selected sample of school-going adolescents (Moussa et al. 2008). Similarly, results from the NSPHUAE show that 1.5% (95% CI 1.0–2.1) of males compared to 0.5% (95% CI 0.3–0.8) of females reported a self-diagnosis of diabetes ($p < 0.001$) (Barakat et al. 2021).

Conversely, a strong female preponderance among adolescents with a diabetes diagnosis has been found in five studies in this review. Two studies were from Saudi Arabia (Al-Ghamdi and Fureeh 2018; Habeb et al. 2011), and the rest were from the UAE (Punnose et al. 2005), Kuwait (Shaltout et al. 2017) and Egypt (Ali et al. 2013). Two studies found the prevalence of T2D to be higher among females. Specifically, Punnose et al. (2005) found that among adolescents from the UAE, the

**Table 10.2** Association of risk factors and characteristics for T1D and T2D among adolescents residing in the MENA region

| Author/year | Risk factors and characteristics |
|---|---|
| Al Amiri et al. (2015) | Adiposity, family history of T2D, employment and high levels of triglycerides were risk factors associated with abnormal glycemic testing<br>Abnormal glycemic testing was 1.9 times more common among adolescents with a first-degree relative and 2.28 times more common among those with high levels of triglycerides |
| Al-Daghri et al. (2015) | Vitamin D deficiency was significantly associated with T2D [OR 3.47 (CI 1.26–5.55); $p < 0.05$] and pre-DM [OR 2.47 (CI 1.48–4.12); $p < 0.01$] in males |
| Ali et al. (2013) | There were significantly more females than males with T2D (64.3% vs. 58.2%)<br>T2D patients had a highly significant higher waist circumference than T1D patients<br>Obesity, female gender, and a positive family history were risk factors for T2D |
| Osman et al. (2013) | 35 (92.1%) had T2D onset between 11 and 18 years of age and were all pubertal<br>Thirty-five (92%) had family history of T2D<br>Twenty-nine (76.3%) of the cases were obese, 8 (21.1%) overweight and only one (2.6%) had normal BMI |
| Al-Agha et al. (2012) | Among adolescents with T2D, 62.86% had a body mass index BMI ≥ 85th percentile, and 37.14% had a BMI ≥ 95th percentile<br>Mean, SD, and median age of diagnosis: 13.1 ± 2.02 and 13.5 years<br>60% were pubertal |
| Habeb et al. (2011) | The incidence was significantly higher in the 10–12-year age group (46.5 per 100,000; 95% CI 38.9–55.2) than in younger children ($p < 0.001$)<br>Incidence rate was significantly higher in females (33.0 per 100,000, 95% CI 29.1–37.3) than males (22.2 per 100,000, 95% CI 19.1–25.7; $p < 0.001$)<br>More cases of T1D were diagnosed during autumn and winter months (249; $p = 0.002$) than during spring and summer (March–August: 170, 40.6%) |
| Cherian et al. (2010) | No etiological influences of maternal age at birth, birth order, birth weight, early introduction of cow's milk and cereals, infections, and vaccines as well as nitrate levels in drinking water were noted in 119 adolescents (<15 years) with T1D between 1980 and 2009<br>An association with cool months (November–February) was found for an increased incidence of T1D during 1990–1994 and 1995–1999 |
| Moadab et al. (2010) | 9.34% ($n = 706$) were overweight and 5.3% ($n = 403$) were obese<br>Impaired glucose tolerance and insulin resistance were detected in three and six participants with IFG, who consisted 0.4% and 0.8% of total obese and overweight students, respectively |
| Moussa et al. (2008) | Significant difference in prevalence between males (47.3, 95% CI 28.7–65.8) and females (26.3, 95% CI 14.8–37.8) at $p = 0.05$<br>Significant trend for an increase in prevalence of T2D with age ($p = 0.026$). 6–9 (17.2); 10–13 (31.6); 14–18 (46.8) per 100,000<br>Adolescents with T2D had a significantly higher frequency (51.1%) of a positive family history of diabetes than those of a similar age without T2D (22.2%) ($p = 0.004$) |

(continued)

**Table 10.2** (continued)

| Author/year | Risk factors and characteristics |
|---|---|
| Moussa et al. (2005) | Family history of T1D and T2D and thyroid disease was associated with T1D based on a pair-matched case-control study conducted in Kuwait (cases: 348; controls: 348)<br>The risk for adolescents (6–18 years) with positive family history of T1D of becoming diabetic was 2.42 times higher than a child with no family history of diabetes (unadjusted odds ratio = 2.42, 95% CI = 1.62–3.67, $p < 0.001$)<br>The risk for T2D was 2.83 times higher (unadjusted odds ratio = 2.83, 95% CI = 1.72–4.68, $p < 0.001$) |
| Punnose et al. (2005) | 10 of the 11 patients diagnosed were female (M:F ratio 1:10)<br>The mean age at diagnosis was 14.6 (SD 3.0) years (median 15 years, range 8–18 years)<br>A positive parental history was obtained in 8 patients with 2 of them also having second-degree relatives with T2D<br>The clinical characteristics were pubertal onset, female preponderance, obesity, and strong family history of T2D |
| Punnose et al. (2002) | Four (80%) were obese with a positive family history of Type 2 DM<br>The mean age of onset of T1D: 9.2 ± 4.1 years with the peak incidence being in the 10–14-year age group |
| Shaltout et al. (2002) | Significant seasonal variation in the number of newly diagnoses cases of T1D was found among the total ($p < 0.001$), with an increase in autumn/winter and decrease in summer |

female to male ratio of T2D diagnosis was 10:1. In Egypt, being female was suggested as a risk factor for T2D (Ali et al. 2013). This study found that among individuals diagnosed with T2D (13.3%), there was a significant difference in the prevalence of females (64.3%) diagnosed than their male counterparts (35.7%). In Saudi Arabia, T1D was also more common among females at 57.5% compared to 42.5% among males (Al-Ghamdi and Fureeh 2018). Similarly, an earlier study conducted in Saudi Arabia in Al-Madinah city found that the incidence of T1D was significantly high among females than in males (33.0 vs. 22.2 per 100, 000; $p < 0.001$) (Habeb et al. 2011). In Kuwait, although there were no significant differences, the standardized incidence rate for T1D was reported to be 44.1 for females compared to 39.2 for males (Shaltout et al. 2017).

## Obesity

The steady decrease in age of onset of T2D beginning during adolescence rather than what has been previously considered as a risk factor correlated with adulthood, has been reflected as a consequence of rapid modernization with major lifestyle changes resulting in childhood obesity, and insulin resistance inducing glucose intolerance (Moussa et al. 2008). In this review, obesity was found as a characteristic and as an associated risk factor in five studies for T2D and prediabetes symptoms.

In one study from the UAE, 5.4% and 0.87% of 1034 overweight/obese Emirati students (11–17 years) had T2D and were positive for prediabetes, respectively

(Al-Amiri et al. 2015). A retrospective cross-sectional study conducted on Saudi overweight and obese adolescents (with body mass indexes of ≥85th and 95th percentile) found that the prevalence of T2D was higher at 9.04% in this population (Al-Agha et al. 2012). Among Saudi adolescents in Riyadh who were considered healthy, Al-Daghri et al. (2015) found that both the prevalence of obesity and elevated fasting blood glucose levels (equivalent to prediabetes; 5.6–6.9 mmol/L) were significantly higher among males than females, at 20.5% versus 17.7% ($p < 0.001$) and 17.8% versus 12.4% ($p < 0.003$), respectively. A large cross-sectional study that screened 672 overweight and obese school adolescents from Iran found that prediabetes symptoms (i.e., impaired glucose tolerance and insulin resistance) were detected in 0.4% and 0.8% of total obese and overweight students, and the overall prevalence of IFG was 4.61% (Moadab et al. 2010). The detection and prevalence of impaired glucose tolerance and insulin resistance among obese and overweight adolescents indicate the increased risk of T2D development in this cohort.

In Egypt, among adolescents <18 years, obesity was considered a risk factor for T2D, based on results showing that a significantly higher BMI and waist circumference centiles for age and sex was seen in adolescents with T2D than those with T1D (Ali et al. 2013). A study conducted in Sudan found that the appearance of obesity was also identified in 29 (76.3%) of the adolescent cases with T2D, while 8 (21.1%) were overweight, and only one (2.6%) had normal BMI (Osman et al. 2013). Most of the cases (86.8%) were coming from urban areas, and 55.3% of them were from high social class (Osman et al. 2013). Interestingly, this study also found that very obese children had earlier onset of T2D (8–11 years) in comparison to nonobese adolescents that had onset between 11 and 18 years. A family history of obesity was also found in this cohort.

## Familial Disposition

Diabetes is a highly heritable condition for both T1D and T2D. In particular, literature suggest that 90% of children and youth affected by T2D have a first- or second-degree relative who also have T2D (Panagiotopoulos et al. 2018). Six studies (UAE:3; Egypt:1; Sudan:1; Kuwait:1) in this review reported family history of diabetes as a risk factor for T2D, while one study from Kuwait suggested a significant association for T1D (Al Amiri et al. 2015; Ali et al. 2013; Moussa et al. 2005, 2008; Osman et al. 2013; Punnose et al. 2002, 2005) (Table 2). Among overweight and obese adolescents from the UAE, family history of T2D was significantly associated with risk of abnormal glycemic testing (based on fasting blood glucose) (Al-Amiri et al. 2015). In Khartoum, Sudan, a large retrospective study found that out of adolescents that had T2D, 92% had a family history of T2D (Osman et al. 2013). A study of Kuwaiti adolescents with T1D matched by age and gender to nondiabetic controls suggested a family history of both T2D and T1D as significant associated factors for developing T1D, in which the risk for T1D was 2.42 times higher (95% CI:1.62–3.67) and T2D was 2.83 times higher (95% CI: 1.72–4.68) (Moussa et al. 2005). This study also indicated a 1.86 times higher risk of being diabetic for adolescents with a family history of thyroid disease (95% CI: 1.18–3.28) than those with no family history.

### Other Factors

Few studies have looked at the role of causative environmental factors as possible contributors to the etiology of T1D incidence. One study in particular conducted in Saudi Arabia aimed to evaluate the role of multiple factors [i.e., maternal age at birth, birth order, birth weight, nutritional factors (cow's milk and cereals), infections and vaccines, as well nitrate levels in drinking water] in the rising incidence of T1D (Cherian et al. 2010). Although no significant associations were found, this study suggested a seasonal variation in T1D development in children during cool months (November–February) compared with warm months (June–September) (Cherian et al. 2010). These findings also aligned with an earlier study from Kuwait which indicated a significant seasonal variation in T1D diagnosis with an increase during autumn/winter (Shaltout et al. 2002), and the study conducted in Al Madinah, which found that 54.9% of adolescents were diagnosed with T1D during winter and autumn months (September–February) compared to 40.6% that were diagnosed during spring and summer (March–August) (Habeb et al. 2011). This seasonality effect could be linked to variability in vitamin D exposure and its immunomodulatory mechanisms in the prevention of T1D (Chakhtoura and Azar 2013).

## 10.4 Discussion

Diabetes mellitus is a noncommunicable disease which has become a major global public health issue. Some regions of the world have experienced greater challenges in controlling the spread of this epidemic, due to variations in health resources, and the influence of various environmental, behavioral, and genetic etiologies. The MENA region in particular is considered one of the highest risk regions for this chronic disease, which may be a result of the rapid changing demographic and recent transition from traditional to westernized nutritional practices, and an increase in sedentary behavior (Al Busaidi et al. 2019; Saeedi et al. 2019; Zabetian et al. 2013). In addition to the various environmental and behavioral factors that are widely known to contribute to diabetes prevalence worldwide, the expression of diabetes may be further influenced by certain social and cultural norms practiced throughout the MENA. Young individuals in particular are becoming targets of intervention as they are impacted by the most common forms of diabetes (T1D and T2D); however, many gaps maintain in the knowledge regarding the various etiology behind the rising epidemic of these diabetes forms.

Over the past two decades, there has been a noticeable increase in diabetes among adolescents throughout the region. In this review, studies from the MENA region that reported on the prevalence and risk factors of diabetes were related to

T1D and T2D. Accordingly, this chapter summarized epidemiological studies related to the prevalence, risk factors, or predictors of T1D and T2D among children and adolescents (0–19 years) from countries in the MENA region, including Saudi Arabia, the UAE, Kuwait, Qatar, Egypt, Iran, and Sudan. Studies from Saudi Arabia reported a significant increase in T1D among adolescents from 1990 to 2016, with incidence rates ranging from 18.05 per 100,000/year for adolescents <15 years old to a prevalence rate of 355/per 100,000 for adolescents 0–19 years old, respectively (Abdul-Jabbar et al. 2010; Al-Ghamdi and Fureeh 2018). In addition, incidence of T1D in Kuwaiti adolescents across two decades has been 4.1% annually, which aligns with international data, suggesting an average relative increase of 3–4% per calendar year worldwide (Patterson et al. 2019; Shaltout et al. 2017; Tuomilehto 2013).

Studies on the prevalence of T2D were limited and remain elusive. Nonetheless, these studies represent some of the predominant risk factors that may be associated with this growing prevalence in the MENA. In this review, factors associated with T2D and their characteristics were family history, sex, pubertal onset, vitamin D deficiency, and obesity. Few studies also reported that adolescents with T2D who were defined as obese also had a family history of T2D. Moreover, environmental and geographic related factors such as living in urban versus rural areas may have an effect on the risk of diabetes development, as suggested by the higher prevalence of people living with diabetes in urban areas (Saeedi et al. 2019; Osman et al. 2013). Geographic factors (e.g., harsh desert climate) coupled with cultural reasons in the MENA may contribute to poor vitamin D status and lead to clinical implications associated with T1D (Rasoul et al. 2016). In this review, vitamin D deficiency was more apparent in male adolescents with T2D, who also showed a high prevalence of prediabetes. This is despite research suggesting that Arab females are more likely to be deficient in vitamin D than Arab males, due to less physical activity outdoors, lack of exposure to sunlight, and are less likely to receive supplementation (Al-Daghri et al. 2015). These contradictory findings suggest that the influence of other gender-related behavioral factors, such as lack of physical activity and poor nutritional practices may be strongly correlated to the presentation of prediabetes symptoms among adolescent Arab males. Further research is required to characterize adolescents with T2D and the associated risk factors that may either initiate or accelerate the autoimmune process leading to pancreatic ß cell destruction (Weiss et al. 2017).

In obese adolescents and children, the risk of diabetes expression and transition from prediabetes symptoms to diabetes are faster than compared to adults, with deterioration of β cells occurring at 20–30% per year (Weiss et al. 2017). Accordingly, the early expression and onset of diabetes during adolescence provides a unique opportunity of intervention to mitigate the risk of development which is associated with many preventable behavioral practices. Specifically, sedentary behavior, nutritional habits consisting of high carbohydrate foods, and the subsequent development of obesity – a major predisposition for the transition from regular glucose metabolism to insulin resistance.

Moreover, the phenomenon of temporary insulin resistance and peak incidence of pubertal onset associated with T1D and T2D suggest homeostasis in glucose metabolism and insulin secretion are dependent upon healthy behaviors during adolescence, composed of balanced energy intake and expenditure involving health dietary behaviors and physical activity, to mitigate the risk of other predisposing factors. In addition, pubertal onset may also differentiate age of diagnosis for sex-related prevalence, as a reflection of differences associated with earlier onset of puberty in females than males (Habeb et al. 2011). In this case, it is critical to examine the earlier onset of symptoms in adolescent females, which may be accelerated in genetically susceptible individuals due to gender-related behaviors, and especially given the higher prevalence of both T1D and T2D among female adolescents found in this review.

The MENA region is composed mainly of Arab populations, which are considered a high-risk population for the development of T2D and T1D. This high-risk consideration stems from a range of epigenetic and genetic factors. Specifically, risk factors for the development of T2D in children and adolescents have been linked to a history of T2D in a first-or second-degree relative, which can also be further expressed throughout generations due to the high prevalence of consanguineous marriages in the region (Al-Amiri et al. 2015; Panagiotopoulos et al. 2018). The prevalence of these sociocultural norms (i.e., preservation of familial structure from unions formed between biologically related individuals) may also be associated with genetic dimorphisms resulting in the expression of risk alleles for T1D (Zayed 2016). The varied expression and contribution of genetic profiles of specific alleles leading to increased incidence of T1D in certain populations may explain the geographic variability in prevalence. This expression may be further initiated or accelerated by the effect of behavioral risk factors. However, further research is required to examine the underlying risk factors and epigenetic mechanisms that may be linked to this difference.

Studies have also illustrated the increase in obesity which has largely been linked to the development of T2D and as a risk factor for the development of gestational diabetes in childbearing women (15–45 years) in the MENA, which further compounds the intergenerational risk of diabetes in offspring (Al-Rifai et al. 2019; Al-Rubeaan, 2015). In fact, in the MENA region, 1 in 9 live births are affected by hyperglycemia in pregnancy (IDF 2019). The same predisposing risk factors have been shared among adolescents, and may be differentiated by gender-related social factors. For example, females face higher risk of sedentary lifestyles, as studies show that factors such as wearing sports clothing, safety concerns, or exercising outdoors for religious reasons are cultural barriers that affect females more than males from Arab nations (Barakat and Yousufzai 2020), which may indirectly contribute to the prevalence of overweight and obesity, and in translation increase the risk of developing T2D.

Overall, there are a limited number of studies reporting on the incidence of adolescent diabetes. In addition, although the countries of the MENA represent some of the highest numbers of T1D, overall this number may be underestimated as only 57% have incidence rates available from the MENA region reported by the IDF (Patterson et al. 2019). Multiple studies suggest the dire need of having a registry in each country with systemic coordination (Alassaf et al. 2019; Zayed 2016), as well as having research done in community-based settings, to further characterize the etiology, mortality, and complication of this serious disease (Saraswathi et al. 2019). Furthermore, multiple characteristics seem to be associated with early onset and diagnosis of diabetes in adolescents, and significant differences are seen between sexes and in relation to gender-related risk factors. Efforts need to be initiated to address prevention strategies and modifiable risk factors of diabetes in children and adolescents and to determine links between social, demographic, and economic contextual factors in developing countries.

## Reflection Questions

1. What are the different forms of diabetes that may affect adolescents and how do they differentiate?
2. What countries have the highest prevalence rates of type 1 diabetes among adolescents in the MENA region?
3. What is the various risk factors that are prominent in the MENA region which may lead to the increased expression of diabetes during adolescence?
4. When do symptoms of type 1 diabetes become apparent among adolescents?
5. What is the key modifiable risk factors that can decrease the prevalence of diabetes and promote long-term positive health?
6. How can policies be tailored to ensure that all countries in the MENA are able to develop registries for diabetes, and why are registries important?

## References

Abduljabbar, M.A., Aljubeh, J.M., Amalraj, A. and Cherian, M.P. (2010). Incidence trends of childhood type 1 diabetes in eastern Saudi Arabia. *Saudi Med J*, *31*(4), pp.413–418.

Abuyassin, B. and Laher, I. (2016). Diabetes epidemic sweeping the Arab world. *World journal of diabetes*, *7*(8), p.165.

Al-Amiri, E., Abdullatif, M., Abdulle, A., Al Bitar, N., Afandi, E.Z., Parish, M. and Darwiche, G. (2015). The prevalence, risk factors, and screening measure for prediabetes and diabetes among Emirati overweight/obese children and adolescents. *BMC Public Health*, *15*(1), pp.1–9.

Al Busaidi, N., Shanmugam, P. and Manoharan, D. (2019). Diabetes in the middle East: government health care policies and strategies that address the growing diabetes prevalence in the middle East. *Current Diabetes Reports*, *19*(2), p.8.

Al Makadma, A.S. (2017). Adolescent health and health care in the Arab Gulf countries: today's needs and tomorrow's challenges. *International Journal of Pediatrics and Adolescent Medicine*, *4*(1), pp.1–8.

Al-Agha, A., Ocheltree, A. and Shata, N. (2012). Prevalence of hyperinsulinism, type 2 diabetes mellitus and metabolic syndrome among Saudi overweight and obese pediatric patients. *Minerva Pediatr*, *64*(6), pp.623–31.

Alassaf, A., Odeh, R., Gharaibeh, L., Ibrahim, S. and Ajlouni, K. (2019). Personal and Clinical Predictors of Poor Metabolic Control in Children with Type 1 Diabetes in Jordan. *Journal of diabetes research, 2019.*

Al-Daghri, N.M., Al-Saleh, Y., Aljohani, N., Alokail, M., Al-Attas, O., Alnaami, A.M., Sabico, S., Alsulaimani, M., Al-Harbi, M., Alfawaz, H. and Chrousos, G.P. (2015). Vitamin D deficiency and cardiometabolic risks: a juxtaposition of Arab adolescents and adults. *PloS one*, *10*(7), p. e0131315.

Al-Ghamdi, A.H. and Fureeh, A.A. (2018). Prevalence and clinical presentation at the onset of type 1 diabetes mellitus among children and adolescents in AL-Baha region, Saudi Arabia. *Journal of Pediatric Endocrinology and Metabolism*, *31*(3), pp.269–273.

Al-Herbish, A.S., El-Mouzan, M.I., Al-Salloum, A.A., Al-Qurachi, M.M. and Al-Omar, A.A. (2008). Prevalence of type 1 diabetes mellitus in Saudi Arabian children and adolescents. *Saudi Med J*, *29*(9), pp.1285–8.

Ali, B.A., Abdallah, S.T., Abdallah, A.M. and Hussein, M.M. (2013). The Frequency of Type 2 Diabetes Mellitus among Diabetic Children in El Minia Governorate, Egypt. *Sultan Qaboos University Medical Journal*, *13*(3), p.399.

Al-Rifai, R.H., Majeed, M., Qambar, M.A., Ibrahim, A., AlYammahi, K.M. and Aziz, F. (2019). Type 2 diabetes and pre-diabetes mellitus: a systematic review and meta-analysis of prevalence studies in women of childbearing age in the Middle East and North Africa, 2000–2018. *Systematic reviews*, *8*(1), p.268.

Al-Rubeaan, K. (2015). National surveillance for type 1, type 2 diabetes and prediabetes among children and adolescents: a population-based study (SAUDI-DM). *J Epidemiol Community Health*, *69*(11), pp.1045–1051.

Alyafei, F., Soliman, A., Alkhalaf, F., Sabt, A., De Sanctis, V., Waseef, R. and Elsayed, N. (2018). Incidence of type 1 and type 2 diabetes, between 2012-2016, among children and adolescents in Qatar. *Acta Bio Medica: Atenei Parmensis*, *89*(Suppl 5), p.7.

Aminzadeh, M., Navidi, N., Valavi, E. and Aletayeb, S.M.H. (2019). Childhood onset type 1 diabetes at a tertiary hospital in south-western Iran during 2000–2015: Rapid increase in admissions and high prevalence of DKA at diagnosis. *Primary care diabetes*, *13*(1), pp.43–48.

Arslanian, S.A. (2000). Type 2 Diabetes Mellitus in Children: Pathophysiology and Risk Factors: Division of Pediatric Endocrinology, Metabolism and Diabetes Mellitus Children's Hospital of Pittsburgh, Pennsylvania, USA. *Journal of Pediatric endocrinology and Metabolism*, *13*(Supplement), pp.1385–1394.

Barakat C., Yousufzai, S. (2020). Health-Risk Behaviors of Adolescents from Arab Nations. In: Laher I. (eds) Handbook of Healthcare in the Arab World. Springer, Cham. https://doi.org/10.1007/978-3-319-74365-3_24-1.

Barakat, C., Yousufzai, S. J., Booth, A., & Benova, L. (2021). Prevalence of and risk factors for diabetes mellitus in the school-attending adolescent population of the United Arab Emirates: a large cross-sectional study. *BMJ open*, *11*(9), e046956.

Buchbinder, M.H., Detzer, M.J., Welsch, R.L., Christiano, A.S., Patashnick, J.L. and Rich, M. (2005). Assessing adolescents with insulin-dependent diabetes mellitus: a multiple perspective pilot study using visual illness narratives and interviews. *Journal of Adolescent Health*, *36*(1), pp.71–e9.

Chakhtoura, M. and Azar, S.T. (2013). The role of vitamin d deficiency in the incidence, progression, and complications of type 1 diabetes mellitus. *International journal of endocrinology, 2013*.

Cherian, M.P., Al-Kanani, K.A., Al Qahtani, S.S., Yesurathinam, H., Mathew, A.A., Thomas, V.S., Mathew, A.M. and Abduljabbar, M.A. (2010). The rising incidence of type 1 diabetes mellitus and the role of environmental factors-three-decade experience in a primary care health center in Saudi Arabia. *Journal of pediatric endocrinology and metabolism*, *23*(7), pp.685–696.

Elissa, K., Bratt, E.L., Axelsson, Å.B., Khatib, S. and Sparud-Lundin, C. (2017). Societal norms and conditions and their influence on daily life in children with type 1 diabetes in the West Bank in Palestine. *Journal of Pediatric Nursing*, *33*, pp.16–22.

Farsani, S.F., Van Der Aa, M.P., Van Der Vorst, M.M.J., Knibbe, C.A.J. and De Boer, A. (2013). Global trends in the incidence and prevalence of type 2 diabetes in children and adolescents:

a systematic review and evaluation of methodological approaches. *Diabetologia*, *56*(7), pp.1471–1488.

Guariguata L, Jeyaseelan, S (2019) Children and Non-communicable Disease – Global Burden Report 2019. Available via http://www.ncdchild.org/media/1603/ncdchild_global_burden-report-2019.pdf. Accessed 20 Nov 2020.

Habeb, A.M., Al-Magamsi, M.S., Halabi, S., Eid, I.M., Shalaby, S. and Bakoush, O., 2011. High incidence of childhood type 1 diabetes in Al-Madinah, North West Saudi Arabia (2004–2009). *Pediatric diabetes*, *12*(8), pp.676–681.

International Diabetes Federation. *IDF diabetes atlas:* (2019). Retrieved from, https://diabetesatlas.org/en/resources/. Accessed 20 Oct 2020.

Kelly, L.A., Lane, C.J., Weigensberg, M.J., Toledo-Corral, C.M. and Goran, M.I. (2011). Pubertal changes of insulin sensitivity, acute insulin response, and β-cell function in overweight Latino youth. *The Journal of pediatrics*, *158*(3), pp.442–446.

King, H., Aubert, R.E. and Herman, W.H. (1998). Global burden of diabetes, 1995–2025: prevalence, numerical estimates, and projections. *Diabetes care*, *21*(9), pp.1414–1431.

Moadab, M.H., Kelishadi, R., Hashemipour, M., Amini, M. and Poursafa, P. (2010). The prevalence of impaired fasting glucose and type 2 diabetes in a population-based sample of overweight/obese children in the Middle East. *Pediatric diabetes*, *11*(2), pp.101–106.

Moussa, M.A.A., Alsaeid, M., Refai, T.M.K., Abdella, N., Al-Sheikh, N. and Gomez, J.E. (2005). Factors associated with type 1 diabetes in Kuwaiti children. *Acta diabetologica*, *42*(3), pp.129–137.

Moussa, M.A., Alsaeid, M., Abdella, N., Refai, T.M., Al-Sheikh, N. and Gomez, J.E. (2008). Prevalence of type 2 diabetes mellitus among Kuwaiti children and adolescents. *Medical Principles and Practice*, *17*(4), pp.270–275.

Osman, H.A., Elsadek, N. and Abdullah, M.A. (2013). Type 2 diabetes in Sudanese children and adolescents. *Sudanese journal of paediatrics*, *13*(2), p.17.

Panagiotopoulos, C., Hadjiyannakis, S. and Henderson, M. (2018). Type 2 diabetes in children and adolescents. *Canadian journal of diabetes*, *42*, pp. S247–S254.

Patterson, C.C., Karuranga, S., Salpea, P., Saeedi, P., Dahlquist, G., Soltesz, G. and Ogle, G.D. (2019). Worldwide estimates of incidence, prevalence and mortality of type 1 diabetes in children and adolescents: Results from the International Diabetes Federation Diabetes Atlas. *Diabetes research and clinical practice*, *157*, p.107842.

Punnose, J., Agarwal, M.M., El Khadir, A., Devadas, K. and Mugamer, I.T. (2002). Childhood and adolescent diabetes mellitus in Arabs residing in the United Arab Emirates. *Diabetes research and clinical practice*, *55*(1), pp.29–33.

Punnose, J., Agarwal, M.M. and Bin Uthman, S. (2005). Type 2 diabetes mellitus among children and adolescents in Al-Ain: a case series. *EMHJ-Eastern Mediterranean Health Journal, 11 (4), 788–797, 2005.*

Rasoul, M.A., Al-Mahdi, M., Al-Kandari, H., Dhaunsi, G.S. and Haider, M.Z. (2016). Low serum vitamin-D status is associated with high prevalence and early onset of type-1 diabetes mellitus in Kuwaiti children. *BMC pediatrics*, *16*(1), p.95.

Reinehr, T. (2013). Type 2 diabetes mellitus in children and adolescents. *World journal of diabetes*, *4*(6), p.270.

Robert, A.A., Al-Dawish, A., Mujammami, M. and Dawish, M.A.A. (2018). Type 1 diabetes mellitus in Saudi Arabia: a soaring epidemic. *International journal of pediatrics, 2018.*

Saeedi, P., Petersohn, I., Salpea, P., Malanda, B., Karuranga, S., Unwin, N., Colagiuri, S., Guariguata, L., Motala, A.A., Ogurtsova, K. and Shaw, J.E. (2019). Global and regional diabetes prevalence estimates for 2019 and projections for 2030 and 2045: Results from the International Diabetes Federation Diabetes Atlas. *Diabetes research and clinical practice*, *157*, p.107843.

Saraswathi, S., Al-Khawaga, S., Elkum, N. and Hussain, K. (2019). A Systematic Review of Childhood Diabetes Research in the Middle East Region. *Frontiers in Endocrinology*, *10*, p.805.

Shaltout, A.A., Qabazard, M.A., Abdella, N.A., LaPorte, R.E., Al Arouj, M., Nekhi, A.B., Moussa, M.A. and Al Khawari, M.A. (1995) High incidence of childhood-onset IDDM in Kuwait. *Diabetes Care*, *18*(7), pp.923–927.

Shaltout, A.A., Moussa, M.A.A., Qabazard, M., Abdella, N., Karvonen, M., Al-Khawari, M., Al-Arouj, M., Al-Nakhi, A., Tuomilehto, J., El-Gammal, A. and Kuwait Diabetes Study Group. (2002). Further evidence for the rising incidence of childhood Type 1 diabetes in Kuwait. *Diabetic medicine*, *19*(6), pp.522–525.

Shaltout, A.A., Wake, D., Thanaraj, T.A., Omar, D.M., Al-AbdulRazzaq, D., Channanath, A., AlKandari, H., Abdulrasoul, M., Miller, S., Conway, N. and Tuomilehto, J. (2017). Incidence of type 1 diabetes has doubled in Kuwaiti children 0-14 years over the last 20 years. *Pediatric diabetes*, *18*(8), pp.761–766.

Taha, T.H., Moussa, M.A.A., Rashid, A.R. and Fenech, F.F. (1983). Diabetes mellitus in Kuwait incidence in the first 29 years of life. *Diabetologia*, *25*(4), pp.306–308.

Tuomilehto, J. (2013). The emerging global epidemic of type 1 diabetes. *Current diabetes reports*, *13*(6), pp.795–804.

Weiss, R., Santoro, N., Giannini, C., Galderisi, A., Umano, G.R. and Caprio, S. (2017). Prediabetes in youths: mechanisms and biomarkers. *The lancet child & adolescent health*, *1*(3), pp.240–248.

Whiting DR, Guariguata L, Weil C, Shaw J. (2011) IDF diabetes atlas: global estimates of the prevalence of diabetes for 2011 and 2030. *Diabetes Res Clin Pract*; 94:311–21.

WHO (2020). Adolescent Health. https://www.who.int/southeastasia/health-topics/adolescent-health. Accessed Nov 2020.

World Health Organization. Global report on diabetes. (2016). https://apps.who.int/iris/bitstream/handle/10665/204871/9789241565257_eng.pdf;%20jsessionid=B27DA6B4FB2DCD29CA71FB7C373A17FA?sequence=1. Accessed 30 Sept 2020.

Zabetian, A., Keli, H.M., Echouffo-Tcheugui, J.B., Narayan, K.V. and Ali, M.K. (2013). Diabetes in the middle East and north Africa. *Diabetes research and clinical practice*, *101*(2), pp.106–122.

Zayed, H. (2016). Epidemiology of diabetic ketoacidosis in Arab patients with type 1 diabetes: a systematic review. *International Journal of Clinical Practice*, 70(3), 186–*195*. doi:https://doi.org/10.1111/ijcp.12777.

Zimmet, P.Z., Magliano, D.J., Herman, W.H. and Shaw, J.E. (2014). Diabetes: a 21st century challenge. *The lancet Diabetes & endocrinology*, *2*(1), pp.56–64.

# Chapter 11
# HIV and Sexual Health in MENA's Adolescents

**Ismaël Maatouk and Moubadda Assi**

## 11.1 Introduction

*Human immunodeficiency virus* (HIV) is a virus that attacks immune cells, making a person more vulnerable to infections and cancers. If left untreated, HIV can lead to the *acquired immunodeficiency syndrome* (AIDS).

The global impact of the HIV and AIDS epidemic has spurred its rapid prioritization as one of the most pressing health issues facing the world community. Although the epidemiology and social forces affecting its continued proliferation differ between communities and regions, HIV continues to spread worldwide. There are approximately 38 million people living with HIV (PLHIV) globally in 2019 and more than 30 million who have died of AIDS-related causes since the beginning of the epidemic (UNAIDS 2020). Global efforts to control the epidemic have been beneficial, and significant progress has been made. For instance, the number of

Biographical statement: Ismaël Maatouk conceived and drafted the paper, Moubadda Assi reviewed all epidemiological facts and data in the manuscript. Ismaël Maatouk and Moubadda Assi approved the final version of the chapter.

I. Maatouk (✉)
Dermatology Department, Clemenceau Medical Center affiliated with Johns Hopkins, Beirut, Lebanon

School of Social Sciences, Nottingham Trent University, London, UK

M. Assi (✉)
School of Social Sciences, Nottingham Trent University, London, UK



C. Barakat et al. (eds.), *Adolescent Health in the Middle East and North Africa*, Global Perspectives on Health Geography,
https://doi.org/10.1007/978-3-030-92107-1_11

newly infected people, including children, and the number of AIDS-related deaths have significantly declined over the years. Moreover, the number of people receiving antiretroviral treatment (ART) increased to 25.4 million in 2019, up from 6.4 million in 2009 (UNAIDS 2020). Under Sustainable Development Goal 3, the global community has committed to end the AIDS epidemic by 2030. Under the "90-90-90" targets, countries work toward achieving, by 2020, "90% of people living with HIV knowing their HIV status; 90% of people who know their HIV-positive status on treatment; and 90% of people on treatment with suppressed viral loads" (United Nations 2015).

Hence, despite significant decrease in incidence, new infections still occur at a high rate, and PLHIV continue to face challenges managing their illness. This is particularly true in the Middle East and North Africa (MENA) region,[1] which witnessed a 22% increase in new infections between 2010 and 2019, despite trends of incidence reduction across all other regions of the world. With respect to the epidemic, countries of the MENA can be categorized into two main groups: countries with a generalized epidemic (i.e., Djibouti, South Sudan, and Somalia) and countries with a concentrated epidemic and low HIV prevalence in the general population (i.e., Lebanon and Egypt). The epidemic is however dynamically growing and still presenting a risk of rapid HIV spread in different pockets of the population (Mumtaz et al. 2020a, b). Moreover, this region has specific characteristics negatively impacting HIV. First, the young population (defined as 15–24 years old) constitutes one third of the regional population (United Nations 2017). Second, at least eight of the MENA countries were classified as being conflict-affected in 2018 (Djibouti, Iraq, Lebanon, Libya, the State of Palestine, Sudan, Syria, Yemen) (The World Bank 2018). Third, the unstable political situation has resulted in millions of mobile or displaced populations, including migrants who might be challenged by risky behavior and might face limited access to health services (the number of international migrants in the MENA surged from 18 million in 2000 to 41 million in 2017 (DESA, U. 2017)). Moreover, the traditionally conservative sexual mores and religious adherence may have played an active role in dissuading populations from engaging in risky behaviors but are waning today.

In addition, there is a growing concern surrounding the potential impact of the coronavirus disease 2019 (COVID-19) pandemic on the HIV response in the MENA, including the availability of ART and access to preventive services. As of October 14, 2020, the region had reported a total of 2,660,450 COVID-19 cases (7% of the global count) with 67,750 deaths (World Health Organization 2020). Disruptions in provision of regular HIV/AIDS services are estimated to result in many more additional deaths (UNAIDS 2020; Jiang et al. 2020).

The purpose of this chapter is to overview the HIV situation among adolescents in the MENA region in terms of the general epidemiology, burden, modes of transmission, preventive and diagnostic tools, management plan, and challenges. We

[1]According to UNAIDS: Algeria, Bahrain, Djibouti, Egypt, Iran, Iraq, Jordon, Kuwait, Lebanon, Libya, Morocco, Occupied Palestinian Territory, Oman, Qatar, Saudi Arabia, Somalia, Sudan, Syrian Arab Republic, Tunisia, United Arab Emirates, and Yemen.

therefore conducted an extensive review of literature articles and other reports from the MENA to include them in our analysis. Our chapter will also discuss some other HIV-related issues from the wider angle of teenagers' sexual health.

## 11.2 Epidemiology and Burden of HIV and STIs in Adolescents in the MENA

By the end of 2019, there were approximately 240,000 PLHIV in the MENA, including 9600 children aged 0–14 (UNAIDS 2020). The region counted 8000 adult and child deaths due to AIDS. The number of new HIV infections increased by 22% in 2019 compared to 2010, whereas AIDS-related deaths decreased by only 2% in 2019 compared to 2010. The percent of PLHIV aware of their status in 2019 was 52%. PLHIV who were on ART reached 38% and PLHIV who were virally suppressed reached 32%.

In the MENA, with the exception of Djibouti, Somalia, and South Sudan, HIV prevalence among the general population is well below 1% (Mumtaz et al. 2020a, b). However, 97% of the new infections are affecting key populations: people who inject drugs (PWID), men who have sex with men (MSM), and female sex workers (FSW) (Mumtaz et al. 2020a, b). Among PWID, there are registered HIV epidemics in Libya (40.7%), Iran (9.4%), and Pakistan (8%) and possibly emerging epidemics in Egypt and Afghanistan. Among MSM, Iran (14.8%), Lebanon (12.3%), Tunisia (10.1%), Sudan (7.8%), Pakistan (7.5%), Egypt (6.2%), Yemen (5.9%), and Morocco (4.9%) have registered concentrated epidemics. Among FSW, there are concentrated epidemics in South of Sudan (37.9%), Libya (15.7%), Sudan (7.7%), Somalia (5.2%), and Morocco (5.1%) (UNAIDS 2020).

According to UNAIDS, there are 6300 adolescents (aged 10–19) living with HIV in the MENA (increase of 28% compared with adolescents living with HIV in 2010). However, the number of new HIV infections across this age category is constantly <1000 per year.

Mother-to-child transmission (MTCT) is defined as the transmission of HIV from an HIV-positive mother to her child during pregnancy, labor, delivery, or breastfeeding. In the absence of any intervention (ART), transmission rates range from 15% to 45%. This rate can be reduced to below 5% with effective ART, leading to undetectable viral loads in the maternal circulation. The MENA has an alarming rate of MTCT of 30%, the highest among all regions, reaching almost three times the global rate (11%). This can be explained by the lack of access to ART, since only 29% of pregnant women needing ART in 2019 were actually able to receive it.

Data on sexually transmitted infections (STIs) are scarce, and available data is often lacking granular age disaggregation. The limited available data, however, reveals that STIs are more common among young people than among other age groups.

A 2007 study among married women in Oman found that age was the most important risk factor for STIs. Women under age 25 were twice as likely to have an STI compared to women who were 25 and older (DeJong et al. 2007). Consistently, a study from Lebanon among gay and bisexual men showed in 2019 that the mean age of the first sexual experience among participants was 16.5 (Maatouk et al. 2019).

Thus, these scarce findings should set an agenda to further explore STIs among adolescents in the MENA and to further determine their prevalence among general and key populations, especially that the first sexual contacts appear to happen during adolescence.

## 11.3 HIV Prevention and Diagnostic Tools

Effective prevention in the MENA depends on reaching and engaging with key populations. For instance, Lebanon, Iran, and Morocco have expanded opioid substitution therapy (OST) programs at the levels of communities and prisons. Consequently, the three countries have reported access to sterile injecting equipment by more than 70% of people who inject drugs. These countries serve as an example where access to OST and sterile injecting equipment was rapidly scaled up in major urban centers and can be helpful for implementation in prevalent countries such as Algeria, Egypt, Lebanon, and Tunisia (UNAIDS 2015).

In relation with MSM programs, Lebanon is an example of success, with 75% of MSM in Beirut reporting both knowledge of their HIV status and condom use. In fact, such programs can be successful when rights, privacy, and dignities of beneficiaries are respected and when the civil society and the key affected communities are implicated in the response to ensure efficient service delivery. Morocco and Tunisia are on the same way, but testing coverage and condom use were still below 50% in 2015. Consistently, pre-exposure prophylaxis usage by MSM communities in Algeria, Iran, Lebanon, and Morocco is still in the first steps of implementation.

In 2018, two new policies were recommended by the WHO: HIV self-test (HIVST) and partner notification (PN) (WHO 2016).

- HIVST is an example of a self-care intervention that recognizes individuals as active agents in managing their own health, including disease prevention, self-medication, and providing care to dependent persons. Vulnerable populations such as MSM and refugees or migrants may avoid the health system due to stigma from providers or difficulties in access to health services. Thus, HIVST aims at scaling up HIV testing. Moreover, since the beginning of the COVID-19 pandemic and the consequent lockdown, more consideration was given to HIVST, since it became the only possible diagnostic tool that can easily be used by people who wished to get tested (Maatouk et al. 2020a, b, c).
- PN services need special attention along with a multidimensional support system from health, psychological, religious, social, and legal perspectives. Although existing evidence supports its safety, implementing PN services in the MENA

requires careful planning. This should take into consideration the nature of the target populations, their environment, and sociocultural beliefs in order to address the beneficiaries' concerns, improve acceptance, maximize uptake, and safeguard from adverse consequences. A crucial role can be played by civil society organizations which have the potential to reach greater numbers of people particularly those unlikely to go to a facility for testing. Partners can also be reached using new approaches like HIVST (Maatouk et al. 2019).

Regarding other communities, such as women engaged in transactional sex and FSW, several countries reported success through high rates of condom use among these key populations: Algeria and Lebanon (>80%) and Djibouti, Iran, Jordan, Morocco, and Tunisia (50–80%). Taking into consideration that these communities are hard to reach, condom use results are understandably hard to achieve. In fact, the low access to HIV testing among these groups (compared to HIV testing access among MSM) is another indicator of the difficulty to reach these populations by prevention programs. It is true that in Lebanon, this coverage is considered good; however in Algeria, Morocco, Tunisia, and Iran, the respective figures are less than 33%. Effective prevention intervention should be based on rights and stigma-free to be able to reverse the trajectory of new infections in these communities.

It is true that ART coverage in the MENA is the lowest in the world; however, testing coverage along with ART coverage is rapidly improving in the region. Innovative approaches were recommended and implemented to achieve these targets, such as reducing the time spent on counseling, HIVST, mobile testing services and outreach, and provision of testing services by lay-providers.

Stigma-free civil organizations available in Lebanon, Jordan, and Morocco among other countries serve as an example of good practice of free, voluntary, and confidential testing service. HIVST was introduced in many MENA countries and aims at scaling up testing and reaching those who do not test.

## 11.4 Factors Contributing to HIV Risk for MENA Young People

### *11.4.1 Vulnerable and Most at-Risk Young People*

Young people in general are considered vulnerable. Among them, special subgroups are considered even more vulnerable and include young refugees/migrants, young people with disability, adolescent girls and early-married girls, unemployed young people, and young people with low socioeconomic status. But the most vulnerable population (at the highest risk of HIV) is attributed to young PLHIV, PWID, sex workers, and MSM.

Poverty particularly puts some communities at great risk of contracting HIV. For instance, street children (boys and girls) might be obliged to sell sex amid a setting where access to condom might probably be low. Several reports shed the light on this increasing cohort in the MENA region, particularly among homeless MSM (Abu-Raddad et al. 2010). In Egypt, 65.8% of homeless MSM had their first same-sex encounter before the age of 15 (El Sayyed et al. 2008). Consistently, a study from Lebanon among gay and bisexual men showed in 2019 that the mean age of the first sexual experience among participants was 16.5 (Maatouk et al. 2019). Other studies have also shown that MSM have women partners as well. For instance, a study among MSM from Libya reported that 40% of respondents had sex with both men and women (Valadez et al. 2013). A survey in Morocco among MSM found similar results (Johnston et al. 2013).

Commercial sex work can also have a negative impact on young people at risk of HIV. This is particularly secondary to the high number of partners and the low condom use. It is worth mentioning that commercial sex work is not exclusive to FSW in the MENA but also exists among MSM (Morocco and Somalia) (IOM 2012; Johnston et al. 2013).

While there are few studies about condom use among key populations in the MENA, surveyed communities show an overall inconsistent use of condoms. For instance, among Sudanese FSW surveyed, 30.3% reported condom use at last sex (Elhadi et al. 2013). This is consistently coupled with a low knowledge about HIV (Elhadi et al. 2013; Maatouk et al. 2019). Other contributing factors include risky behaviors and limited access to services (Navadeh et al. 2013).

### *11.4.2 Lack of Sexual and Reproductive Health Education*

The absence of accessible, adequate, and effective sexual and reproductive health education in MENA schools is one of the major gaps that puts young people at risk of contracting HIV.

In 2016, a study of students from 17 universities across Lebanon revealed a low percentage of condom use (36.3%) (Salameh et al. 2016). A very recent study by Maatouk et al. in 2020 showed that females tend to lack self-efficacy in relation to sexual behavior, to be coerced into behaviors they do not desire, and to engage in STI screening less frequently, while men appear to use substances more frequently, which could lead to poor outcomes. Furthermore, on the whole, heterosexual students engage less frequently with sexual health screening than non-heterosexual students.

Many health education interventions assessing medical students' HIV knowledge and attitudes found a positive impact in decreasing stigma and discriminatory attitudes toward PLHIV post-intervention. These studies were conducted in Egypt (El-Nawawy 2008), Libya (Sugathan and Swaysi 2012), the UAE (Barss et al. 2009), and Yemen (Badahdah and Sayem 2010). Thus, sustained educational programs can only have a greater positive impact on access to key HIV services.

According to a systematic review, the overall basic HIV knowledge was high among key populations at higher risk of infection and bridging and general population groups (Mumtaz et al. 2020a, b). However, few population pockets still had low basic knowledge. The level of comprehensive knowledge was overall low, and misinformation and misconceptions were prevalent. At-risk communities such as PWID, MSM, and FSW were unaware of some modes of HIV transmission. Perception of risk of infection was low even among these at-risk communities. Differentials in knowledge were raised putting women, rural populations, refugees, and other marginalized minorities at a disadvantage. Attitudes toward PLHIV tended to be negative.

### *11.4.3 Conflict, Economic Downturn, and Migration*

Conflict and economic downturn lead to decreased HIV services, lack of funds to offer testing and prevention services, lack of sustainability of services, deprioritization of sexual health, and delayed marriages coupled with stress-induced risky behaviors. In addition, unemployment lead to more spare time for young people to socialize, which possibly exposes them to risky behaviors.

Migration and displacement, which happens in the MENA for several political and economic reasons, plays a crucial role in the spread of HIV. The populations affected by migration or displacement definitely include young people (Abu-Raddad et al. 2010). These communities are usually marginalized and lack access to services. Moreover, the increase in sex work for financial purposes makes these communities even more vulnerable to HIV.

### *11.4.4 Gender Disparity*

MENA still ranks in the bottom 20% of the global health gender gap and continues to rank last on the overall gender gap index, behind South Asia (The Global Gender Gap Report 2017).

According to Madani (2018), women are particularly susceptible to the HIV infection. Not only are women more biologically prone to acquiring the infection, but endemic interpersonal violence against women and girls, misogynistic traditions, lack of control over condom use, and limited sex education reinforce unequal power dynamics between men and women. Repeatedly, studies on notified cases have indicated that the vast majority of women who acquire HIV acquire it from their husbands (Abu-Raddad et al. 2010; Mumtaz et al. 2020a, b).

### *11.4.5 Risky Behaviors*

This section will focus on risky behaviors that are specific to some key populations such as PWID, MSM, and FSW.

Sharing of non-sterile needles and/or syringes is one of the key risky behaviors that expose PWID to HIV infection. An average of 42% of people who inject drugs in MENA ever shared needles/syringes, while 24% did so in the last injection (Mumtaz et al. 2020a, b).

Having multiple, including concurrent, sexual partners are reported by over 90% of MSM in the MENA with an average of 4–14 sexual partners in the last 6 months (Mumtaz et al. 2020a, b). Male sex work also appears to be common, with 20–76% of MSM reporting ever exchanging sex for money (Mumtaz et al. 2020a, b). Overall, the rate of consistent condom use is below 25% (Mumtaz et al. 2020a, b).

According to Mumtaz et al. (2020a, b), the mean number of clients of FSW in the past month is in the range of 4.4–114 with a median of 34 clients. Consistent condom use is reported by about a third of all FSW.

### 11.4.6 Lack of Effective Surveillance

Besides the repercussions of political instability on surveillance and reporting, the MENA region is traditionally conservative. This confluence of sociopolitical realities presents serious challenges to effectively assessing HIV prevalence among young populations of the region (Shawky et al. 2009). While a second-generation surveillance has been recommended, the majority of MENA countries rely on passive reporting—with active reporting only in place for pregnant women at antenatal clinics (Shawky et al. 2009). This is also true for non-HIV STIs, such as *Neisseria gonorrhea* (Maatouk and Assi 2020).

### 11.4.7 Lack of Research

The absence of research on risky behavior among young people further contributes to the incomplete baseline of knowledge on HIV/AIDS proliferation factors. Numerous studies exist on MENA young peoples' knowledge of and attitude toward HIV; however, few of them ask behavioral questions. Maatouk et al. (2020a, b, c) reported that Lebanese MSM ($n$ = 1364) who engage in riskier behaviors and who appraise their risk to be high are more likely to get tested for HIV/STIs. Aside from reports usually collected in university hospital settings, it is not expected to have official studies about young people's sexual behavior because this may be perceived as an approval or legitimacy of these behaviors. The truth is the majority of MENA countries stigmatize and criminalize some of these behaviors. This results in the avoidance of collecting such information.

## 11.5 Conclusions

Many factors in the MENA contribute to the spread of HIV among young populations. Youth bulge, increased globalization, ongoing conflict, mass migration, and continued conservative approach to sex and reproductive health education and service access should all be addressed to limit the spread of HIV among MENA's young people. COVID-19 is one of the new challenges added to the already exhausted health systems. The continued efforts to sustain HIV prevention, care, and treatment to all populations in need keep alive the promise of an HIV-free generation and a healthy future for a MENA region still coping with HIV challenges.

## Critical Questions

1. What are the missed opportunities for the Middle East and North Africa region to achieve its HIV goals, especially in the young populations?
2. Which factors contribute to HIV risk for MENA young people?
3. Which behaviors are considered at risk?
4. How does COVID-19 contribute to the existing challenges in the regional HIV situation?
5. How to overcome COVID-19 challenges in HIV response?

## References

Abu-Raddad, L.J., Ayodeji Akala, F., Semini, I. Riedner, G. Wilson D., & Tawil, O., 2010. Characterizing the HIV/AIDS Epidemic in the Middle East and North Africa: Time for Strategic Action. The World Bank.

Badahdah AM, Sayem N, 2010. HIV-related knowledge and AIDS stigma among college students in Yemen. EMHJ-Eastern Mediterranean Health Journal, 16 (8), 901–906.

Barss P, Grivna M, Ganczak M, Bernsen R, Al-Maskari F, El Agab H, Al-Awadhi F, Al-Baloushi H, Al-Dhaheri S, Al-Dhahri J, Al-Jaberi A, 2009. Effects of a rapid peer-based HIV/AIDS educational intervention on knowledge and attitudes of high school students in a high-income Arab country. JAIDS Journal of Acquired Immune Deficiency Syndromes 52:86–98

DeJong J, Shepard B, Roudi-Fahimi F, Ashford L, 2007. Young people's sexual and reproductive health in the Middle East and North Africa. Reprod Health. 14:8.

DESA, U., 2017. Population Division. 2017. Trends in International Migrant Stock: The 2017 Revision. United Nations Database. POP/DB/MIG/Stock/Rev.

El Sayyed, N., Kabbash, I.A. and El Gueniedy, M., 2008. Risk behaviours for HIV/AIDS infection among men who have sex with men in Cairo, Egypt. *EMHJ-Eastern Mediterranean Health Journal, 14 (4), 905–915, 2008.*

Elhadi M, Elbadawi A, Abdelrahman S, Mohammed I, Bozicevic I, Hassan EA, Elmukhtar M, Ahmed S, Abdelraheem MS, Mubarak N, Elsanousi S, 2013. Integrated bio-behavioural HIV surveillance surveys among female sex workers in Sudan, 2011–2012. Sexually transmitted infections. 89(Suppl 3):iii17–22

El-Nawawy, A. N., 2008. The effect of health education on HIV knowledge and attitude of medical students. The Egyptian Journal of Community Medicine, 26(3).
IOM, 2012. Youth Behavioural Survey Report: Somalia 2012.
Jiang H, Zhou Y, Tang W, 2020. Maintaining HIV care during the COVID-19 pandemic. The Lancet HIV. 7(5):e308–9.
Johnston LG, Alami K, El Rhilani MH, Karkouri M, Mellouk O, Abadie A, Rafif N, Ouarsas L, Bennani A, El Omari B, 2013. HIV, syphilis and sexual risk behaviours among men who have sex with men in Agadir and Marrakesh, Morocco. Sexually transmitted infections. 89(Suppl 3):iii45–8
Maatouk I, Assi M, 2020. Assessment of Neisseria gonorrhea prevalence and resistance in the Eastern Mediterranean Region. Dermatol Ther *(in press)*
Maatouk I, Assi M, Hermez J, 2019. Partner notification in the Eastern Mediterranean Region: is there a way? East Mediterr Health J. 25:660–7
Maatouk I, Assi M, Jaspal R, 2020a. How can we enhance sexual health outcomes in men who have sex with men in Lebanon? BMJ Sex Reprod Health *(in press)*
Maatouk I; Nakib M, Assi M, Farah P, Makso B, Nakib C, Rady A, 2020b. Community-led HIV self-testing for men who have sex with men in Lebanon: lessons learned and impact of COVID-19. BMC Supplements (in press)
Maatouk I, Assi M, Jaspal R. 2020c. Predicting sexual risk and sexual health screening in a sample of university students in Lebanon: a cross-sectional study. Eastern Med Health J (in press)
Madani N, 2018. Gender equality is crucial to the fight for better HIV treatment access and outcomes in the MENA region. Journal of the International AIDS Society. 21(3).
Mumtaz GR, Hilmi N, Majed EZ, Abu-Raddad LJ, 2020a. Characterising HIV/AIDS knowledge and attitudes in the Middle East and North Africa: Systematic review and data synthesis. Global Public Health. 15:275–98.
Mumtaz GR, Chemaitelly H, Abu-Raddad LJ, 2020b. The HIV Epidemic in the Middle East and North Africa: Key Lessons. Handbook of Healthcare in the Arab World. Springer Nature Switzerland AG 2020 1 I. Laher (ed.).
Navadeh S, Mirzazadeh A, Gouya MM, Farnia M, Alasvand R, Haghdoost AA, 2013. HIV prevalence and related risk behaviours among prisoners in Iran: results of the national biobehavioural survey, 2009. Sexually transmitted infections. 89(Suppl 3):iii33–6
Salameh, P., Zeenny, R., Salamé, J., Waked, M., Barbour, B., Zeidan, N., & Baldi, I., 2016. Attitudes towards and practice of sexuality among university students in Lebanon. *Journal of Biosocial Science, 48(02), 233–248.*
Shawky S, Soliman C, Kassak KM, Oraby D, El-Khoury D, Kabore I, 2009. HIV surveillance and epidemic profile in the Middle East and North Africa. Journal of acquired immune deficiency syndromes (1999). 51(Suppl 3):S83.
Sugathan S, Swaysi M, 2012. Knowledge about HIV/AIDS among Premedical Students in Misurata, Libya and the Effectiveness of a Health Education Intervention. J Community Med Health Educ. 2:2161–711.
The Global Gender Gap Report 2017. World Economic Forum. Available from: https://www.weforum.org/reports/the-global-gender-gap-report-2017
The World Bank, 2018. Harmonised List of Fragile Situations FY19. Available at: http://pubdocs.worldbank.org/en/892921532529834051/FCSList-FY19-Final.pdf
UNAIDS, 2015. HIV in the Middle East and North Africa 2013 – 2015. Available at: https://open.unaids.org/sites/default/files/documents/HIV%20in%20MENA.pdf
UNAIDS. 2020 Global AIDS Update: Seizing the Moment; July 2020. UNAIDS. AIDS info website; accessed July 2020, available at: http://aidsinfo.unaids.org/. UNAIDS. Core Epidemiology Slides; July 2020.
United Nations, 2015. *Transforming our world: the 2030 Agenda for Sustainable Development.*
United Nations, 2017. World population prospects 2017.

Valadez JJ, Berendes S, Jeffery C, Thomson J, Othman HB, Danon L, Turki AA, Saffialden R, Mirzoyan L, 2013. Filling the knowledge gap: measuring HIV prevalence and risk factors among men who have sex with men and female sex workers in Tripoli, Libya. PLoS One. 8:e66701.

World Health Organization, 2016. HIV/AIDS. Guidelines on HIV Self-Testing and Partner Notification. Available at http://apps.who.int/iris/bitstream/10665/251655/1/9789241549868-eng.pdf

World Health Organization Eastern Mediterranean Regional Office, 2020. Updates on COVID-19. Available at: http://www.emro.who.int/pandemic-epidemic-diseases/covid-19/covid-19-situation-updates-for-week-31-26-july1-august-2020.html

# Index




C. Barakat et al. (eds.), *Adolescent Health in the Middle East and North Africa*, Global Perspectives on Health Geography,
https://doi.org/10.1007/978-3-030-92107-1

The manufacturer's authorised representative in the EU is Springer Nature Customer Service Centre GmbH, Europaplatz 3, 69115 Heidelberg, Germany. If you have any concerns regarding our products, please contact ProductSafety@springernature.com

Printed and bound by CPI Group (UK) Ltd, Croydon, CR0 4YY
02/10/2026
02227130-0002